Masculinities in the Field

Full details of all of our publications can be found on http://www.channelviewpublications.com, or by writing to Channel View Publications, St Nicholas House, 31–34 High Street, Bristol BS1 2AW, UK.

Masculinities in the Field

Tourism and Transdisciplinary Research

Edited by
**Brooke A. Porter,
Heike A. Schänzel and
Joseph M. Cheer**

CHANNEL VIEW PUBLICATIONS
Bristol • Blue Ridge Summit

DOI https://doi.org/10.21832/PORTER7963
Library of Congress Cataloging in Publication Data
A catalog record for this book is available from the Library of Congress.
Names: Porter, Brooke A., editor. | Schänzel, Heike, editor. | Cheer, Joseph M., editor.
Title: Masculinities in the Field: Tourism and Transdisciplinary Research / Edited by Brooke A. Porter, Heike A. Schänzel and Joseph M. Cheer.
Description: Bristol, UK; Blue Ridge Summit, PA: Channel View Publications, 2021. | Includes bibliographical references and index. | Summary: "This volume is an essential reference for designing, analysing and reflecting on field research. It advances the literature on gender by taking a specific focus on masculinity. The accounts of gendered field experiences further the call for gender positionality in research, will aid researchers and be a useful tool for supervisors" — Provided by publisher.
Identifiers: LCCN 2020043689 (print) | LCCN 2020043690 (ebook) | ISBN 9781845417956 (paperback) | ISBN 9781845417963 (hardback) | ISBN 9781845417970 (pdf) | ISBN 9781845417987 (epub) | ISBN 9781845417994 (kindle edition) Subjects: LCSH: Tourism—Social aspects. | Tourism—Sex differences. | Masculinity. Classification: LCC G156.5.S63 M37 2021 (print) | LCC G156.5.S63 (ebook) | DDC 910.811—dc23 LC record available at https://lccn.loc.gov/2020043689
LC ebook record available at https://lccn.loc.gov/2020043690

British Library Cataloguing in Publication Data
A catalogue entry for this book is available from the British Library.

ISBN-13: 978-1-84541-796-3 (hbk)
ISBN-13: 978-1-84541-795-6 (pbk)

Channel View Publications
UK: St Nicholas House, 31–34 High Street, Bristol, BS1 2AW, UK.
USA: NBN, Blue Ridge Summit, PA, USA.

Website: www.channelviewpublications.com
Twitter: Channel_View
Facebook: https://www.facebook.com/channelviewpublications
Blog: www.channelviewpublications.wordpress.com

Copyright © 2021 Brooke A. Porter, Heike A. Schänzel, Joseph M. Cheer and the authors of individual chapters.

All rights reserved. No part of this work may be reproduced in any form or by any means without permission in writing from the publisher.

The policy of Multilingual Matters/Channel View Publications is to use papers that are natural, renewable and recyclable products, made from wood grown in sustainable forests. In the manufacturing process of our books, and to further support our policy, preference is given to printers that have FSC and PEFC Chain of Custody certification. The FSC and/or PEFC logos will appear on those books where full certification has been granted to the printer concerned.

Typeset by Nova Techset Private Limited, Bengaluru and Chennai, India.
Printed and bound in the UK by Short Run Press Ltd.
Printed and bound in the US by NBN.

Contents

Contributors vii

Foreword xiii
Nigel Morgan and Annette Pritchard

Introduction – Issues in the Field: Masculinities in
  Masculine Spaces 1
Brooke A. Porter, Heike A. Schänzel and Joseph M. Cheer

Part 1: Hegemonic and Heteronormative Masculinities

1 It's Okay to Cry: Encouraging Emotional Writing Among
  Male Tourism Scholars 13
  Jack Shepherd

2 When Is a Hegemonic Male Not a Hegemonic Male?
  Personal Reflections of a Tourism(ish) Researcher 31
  Neil Carr

3 Exploring the Expression of the Masculine in Adventure
  Activities: A Personal Reflection 46
  Mark B. Orams

4 Meditations on Masculinity: Encounters in Salty
  Research Spaces 56
  Jacques D. Mahler-Coetzee

Part 2: Performing Heteronormative Masculinities

5 Performing and Negotiating Filipino Masculinities
  in the Field 71
  Richard S. Aquino

6 How Masculinity Creeps In: Awkward Field Encounters
  of a Male Researcher 85
  Can-Seng Ooi

Part 3: Situated Masculinities

7   A Tale of Two Researchers: Masculinity in Cross-cultural Contexts   101
Joseph M. Cheer and Alan A. Lew

8   Gender, Identity and Discomfort: Negotiating Self and Context in the Field   123
Dominic Lapointe

9   Journeying into Yogaland: A Cautionary Tale of a White Guy's Perspectives on Yoga-related Fieldwork in Japan   134
Patrick McCartney

10   A *Mzungu* in Kenya: Dissonant Masculinity and Ethnographic Field Research in Sub-Saharan Africa   153
Gary Lacey

11   Doing Fieldwork in Palestine: Checkpoints, Access Restrictions, Security and Well-being   170
Rami K. Isaac

Part 4: Paternal Masculinities

12   Finding Gender at the Intersection of Family and Field: Family Presences in Sweden   189
Stuart Reid

13   Fatherhood in the Field: Reflections on Kinship, Identity and Ethnographic Research   206
Michael A. Di Giovine

Masculinities in Tourism Research: Implications and Conclusions   228
Joseph M. Cheer, Heike A. Schänzel and Brooke A. Porter

Index   236

Contributors

Editors

Brooke A. Porter is an Associate Professor of Food and Sustainability Studies at the Umbra Institute in Perugia, Italy. Her research focuses on the human dimensions of fisheries and coastal environments. Her current work explores alternative (and fantastic) methods for engaging the public in conservation action. She also researches tourism as an opportunity for entrepreneurship, development and conservation in lesser-developed regions. Brooke has worked in various capacities with NGOs, international aid agencies and educational institutions in Hawaii, and in New Zealand, the Philippines and Eritrea. She currently serves as a scientific adviser to The Coral Triangle Conservancy, a NGO that focuses on reef protection and restoration in the Philippines.

Heike A. Schänzel is an Associate Professor and Postgraduate Programme Leader in International Tourism Management at the Auckland University of Technology, New Zealand. It has taken her 20 years, between having three children, to finish studying and enter academia. She considers herself as a mid-career researcher with a focus on families, children and gender issues in tourism who draws on her own experiences as a mother travelling and researching. Heike's other research interests include: tourist behaviour and experiential travel; sociality in tourism; femininities and masculinities in tourism research; innovative and qualitative research methodologies; and critical theory development in tourism. She is passionate about better understanding family fun (along with the avoidance of conflict) and the facilitation of sociality and meaningful experiences from the perspectives of diverse families within the context of leisure, tourism and hospitality.

Joseph M. Cheer is a Professor in Sustainable Tourism at the Center for Tourism Research, Wakayama University, Japan. He has previously lectured in sustainable tourism at Monash University, Australia, and is a member of the board of the International Geographical Union (IGU) Commission on Tourism and Leisure and Global Change and a steering committee member for Critical Tourism Studies (CTS) Asia Pacific. Joseph's research draws from transdisciplinary perspectives, especially human/economic geography, cultural anthropology and political economy, particularly in the Asia Pacific. As a former practitioner (industry and government), now academic researcher and practising consultant/analyst, his work emphasises resilience building, sustainability and social justice.

Authors

Jack Shepherd is a PhD student in tourism studies at Mid-Sweden University and the associated European Tourism Research Institute (ETOUR). He previously studied History and French at King's College London and Sciences Po Paris, where his research focused on the dynamics of conflict in Bosnia (1992–1995) and France (1940–1944). His doctoral research continues this interest and focuses on the role tourism can play in peacebuilding in the Israeli -Palestinian context. His work often employs creative methodologies such as creative analytic practice and autoethnography.

Neil Carr is a Professor in the Department of Tourism, University of Otago, New Zealand. His research focuses on understanding behaviour within tourism and leisure experiences, with a particular emphasis on children and families, sex and animals. Neil's recent publications include *Tourism and Animal Welfare* (with Don Broom, 2018, CABI), *Wild Animals and Leisure: Rights and Welfare* (with Young, 2018, Routledge) and *Domestic Animals, Humans, and Leisure: Rights, Welfare, and Wellbeing* (with Janette Young, 2018, Routledge).

Mark B. Orams has competed in an around-the-world yacht race and has many trans-ocean sailing experiences. He has been a member of expeditions to the Antarctic, the Amazon, the remote Kermadec Islands and Sumatra. A surfer, sailor, scuba-diver and ocean-paddler, he has a lifetime of experiences exploring and adventuring in remote locations. Through these experiences Mark has developed a curiosity for the motivations, influences and meanings of such experiences for participants and has explored this as a social scientist. His research focus has been in the marine recreation and tourism realms and he is considered one of the pioneering scholars in these areas. Currently Mark serves as the Co-chair of the International Coastal and Marine Tourism Society; he is on the board of a number of adventure and environmental charities and on the editorial boards of the academic journals *Tourism in Marine Environments* and *Coastal Management*.

Jacques D. Mahler-Coetzee is an attorney, senior lecturer in medical jurisprudence at the Nelson Mandela School of Law, South Africa, leadership instructor in the South African Law Society, trans-Atlantic sailor, scuba

DiveMaster and tour guide. He read environmental law for his LLM (University of Cape Town); his doctoral work considers the sustainable regulation of surfing as a vector for tourism and coastal development in Africa. Jacques currently researches aviation/underwater cultural heritage governance for the Algoa Bay Marine Spatial Planning Pilot Project, Port Elizabeth, South Africa. His recent writing includes the collaborative publication 'Working together for our oceans: A marine spatial plan for Algoa Bay, South Africa' (*African Journal of Science*).

Richard S. Aquino is a lecturer at the Auckland University of Technology in New Zealand, where he also obtained his PhD master's degree in tourism. He has worked as a lecturer and on community-based tourism planning projects in the Philippines. Richard's research focuses on sustainable tourism planning and development, social entrepreneurship in tourism, geotourism, and the application of native methodologies in tourism knowledge production.

Can-Seng Ooi is a sociologist and anthropologist. He is also Professor of Cultural and Heritage Tourism at the University of Tasmania (UTAS). Before joining UTAS in late 2016, he lived and worked in Denmark for 20 years. The examples he presents in the chapter are from his research work at Copenhagen Business School. In 2018 the Women Academics in Tourism (WAiT) network recognised him as an 'Awesome Scholar', giving a nod to his scholarship and his contribution to the collegiality of the tourism research network. He was and still is smitten by that award. Can-Seng's website is at www.cansengooi.com.

Patrick McCartney is currently a Research Associate at the Anthropology Institute of Nanzan University, Nagoya, Japan and a Visiting Fellow at the South and South-East Asia Department at the Australian National University, Canberra, Australia. The research in this edition occurred between 2017–2019 while a JSPS postdoctoral fellow at the Graduate School of Global Environmental Studies at Kyoto University, Japan. More broadly, his research agenda builds from a base that includes archaeology, classical philology, historical linguistics, sociology and linguistic anthropology. This chapter is a branch of the Yogascapes in Japan project, which focuses on the politics of imagination and the economics of desire. The platform of his research agenda is political economy and is framed by yoga's relation to wellness tourism, sustainable development, nation branding, soft power, new religious movements and ethno-nationalism. Of particular interest is the unintended consequences that can arise through the myth-making process for both consumers and producers of yoga-inflected lifestyles. Through exploring the creation of demand and the cultivation of affective desire, this project seeks to identify and articulate the ways in which people can become unwittingly entangled across overlapping social worlds by focusing on the shared utopian aspiration to disrupt a perceptibly disenchanted world and make it better through the adoption of various 'yogic lifestyles'.

Alan A. Lew is Professor Emeritus in the Department of Geography, Planning, and Recreation at Northern Arizona University, USA. His academic background covers the fields of human geography, urban planning and tourism studies, with research mostly in east and southeast Asia. He is the founding Editor-in-chief of the *Tourism Geographies* journal (Routledge), and he has written over 100 journal articles and book chapters along with several books, including *Tourism Geography* (3rd edn, 2014, Routledge). Alan is a Fellow of the International Academy for the Study of Tourism and a member of the American Institute of Certified Planners.

Dominic Lapointe is a Professor in the Department of Urban and Tourism Studies at Université du Québec à Montréal (UQAM). He holds the Chaire de recherche sur les dynamiques touristiques et les relations socioterritoriales and leads the Groupe de recherche et d'intervention tourisme territoire et société (GRITTS) at UQAM. He is also the Head Editor of *Téoros*, a tourism studies academic journal. Dominic's work explores the production of tourism space and its role in capitalist system expansion and its biopolitical dimensions. His latest research looks at climate change, social innovations, indigeneity and critical perspective in tourism studies.

Gary Lacey is an adjunct lecturer in postgraduate research in the Graduate Tourism Program at Monash University, Australia. Having been born in Kenya, his main research interests lie in health, poverty alleviation and empowerment issues in sub-Saharan Africa, with a focus on Kenya and Botswana. Gary has authored or co-authored papers and book chapters on philanthropic tourism in Kenya, community-based tourism in Kenya, the empowerment of women through cultural tourism in Botswana, village-based tourism in the Kalahari Desert, poverty tourism and resilience theory, HIV prevention in Kenya and health and parks in Australia and the USA.

Rami K. Isaac was born in Palestine and undertook his undergraduate studies in the Netherlands, his graduate studies in the UK and earned his PhD from the University of Groningen, the Netherlands. He is currently a senior lecturer in tourism at Breda University of Applied Sciences in the Netherlands. In addition, Rami is an Assistant Professor in the Faculty of Tourism and Hotel Management at Bethlehem University, Palestine. His research interests are in the areas of tourism development, critical theory and the political aspects of tourism. He has published numerous articles and book chapters on tourism and political (in)stability, tourism and war, dark tourism, violence and transformational tourism.

Stuart Reid is a doctoral student in the Department of Service Management and Service Studies at Lund University, Sweden. His dissertation research focuses on understanding the practice of lifestyle enterprising. In particular, by taking a social constructionist perspective and using a grounded theory methodology, his research aims to shed light on and contribute to the theorization of the abundant social phenomenon of lifestyle enterprising. Beyond this, his diverse research interests include digitalization, gender, entrepreneurship, higher education, innovation, tourism and critical theory. Stuart also teaches and supervises students at bachelor's and master's levels in subject areas spanning digitalisation, strategic communication, academic writing, research methodology, tourism value creation and tourism innovation.

Michael A. Di Giovine is Associate Professor of Anthropology at West Chester University, USA, the Director of its Museum of Anthropology and Archaeology and Honorary Fellow at the University of Wisconsin-Madison. A former tour operator who runs an annual field school in Perugia, Italy, his research in Europe and Southeast Asia focuses on tourism, pilgrimage, heritage, foodways and religion. Among his publications are *Tourism and the Power of Otherness* (2014, Channel View Publications), *The Seductions of Pilgrimage* (2015, Ashgate) and *Study Abroad and the Quest for an Anti-Tourism Experience* (2020, Lexington). Michael is the Convenor of the American Anthropological Association's Anthropology of Tourism Interest Group and the editor of Lexington Books' series, *The Anthropology of Tourism: Heritage, Mobility and Society*. His website is at www.michaeldigiovine.com.

Illustrator

Yana Wengel is an associate professor at the Joint Tourism College HAICT of Hainan University in China. Yana takes a critical approach to tourism studies, and her interests include volunteer tourism, non-profit tourism, tourism in developing economies, creative methodologies and mountain tourism. Her doctoral research examined the social construction of host–guest experiences in World Wide Opportunities on Organic Farms programme (WWOOFing). She has an interest in creative qualitative tools for data collection and stakeholder engagement. Yana is a co-founder of the LEGO® SERIOUS PLAY® research community.

Foreword

Boys rapidly learn how they should perform masculinity, which is essentially the opposite of what is feminine (Aboim, 2010). Gender is thus a form of practice, a structure of socially- and culturally-based beliefs and practices that create differences between men and women (Pritchard, 2014). As boys become 'real' men, the attributes of 'Western' masculinity become fixed as they 'man up' and 'bottle up' their emotions. Contemporary conceptualisations of masculinity are heavily problematised and in recent years a toxic 'Western' masculinity has been called out as an explanation for male violence, misogyny and sexism, entitlement, aggression and negative health outcomes. Of course, masculinity is not singular, and Raewyn Connell's pioneering work demonstrated how multiple masculinities are shaped by class, race, culture, sexuality, and so forth, often in competition with one another. Clearly, not all masculinities are negative and damaging but hegemonic masculinity (Connell, 1995), which privileges the position of some men over women and over other (often minority groups of) men perpetuates destructive inequalities.

This important book asks the reader to reflect, perhaps for the first time, on how multiple masculinities shape the tourism research process, right the way through from its research topics and fieldwork to its editorial and publishing decisions. As Jack Shepherd argues in Chapter 1, there is a link between the masculinity that teaches men not to talk about emotions and the pre-eminence of the tourism 'malestream' – a value-free and emotionless approach to enquiry in the field, itself dominated by male gatekeepers. While the tourism academy reflects global higher education's structural gender inequalities – where glass ceilings, sticky floors and maternal walls are reflected in women's under-representation in tenured and senior positions and a serious pay gap – it is a field particularly skewed in favour of men (Pritchard & Morgan, 2017).

Recently in tourism studies we have seen women increasingly reflecting on their positionality as researchers and considering how their gender shapes and impacts the research context and its fieldwork. Indeed, there is now a strong cadre of female tourism scholars whose doctoral theses contain insightful and astute reflections on positionality and femininity in the field. We less commonly encounter men who have committed to text their reflections on their masculinity and its influence on their roles as researchers. Mark Orams notes here (Chapter 3) that in some ways this is surprising because as researchers we are all taught to consider our potential biases and either to seek to minimise them or to be explicit about them (Roulston & Shelton, 2015). In other ways of course, this lacuna is unsurprising as rarely does the hegemonic male (see Neil Carr's chapter) recognise his positionality, since he is the 'norm' against which others are benchmarked. For many senior tourism scholars, in particular, their identities remain unarticulated because as white, middle class men they are the self, the same, the norm against which others are measured, they have 'no class, no race, no gender ... [they are] the generic person' (Kimmel, 1996: 4).

This book maps new territory in tourism studies by foregrounding male reflexivity and is the ideal companion to Brooke Porter and Heike Schänzel's (2018) edited collection, *Femininities in the Field: Tourism and Transdisciplinary Research*. That book analysed and reflected on the effect of femininities in the field and the encountered biases specific to women researchers in tourism studies. Just like that volume, this collection provides case studies using reflexivity to create baselines for comparison for male and female researchers through a transdisciplinary approach in a global context. Its contributors discuss how masculinity shapes every aspect of research, from the topics investigated, the opportunities to gain access to male-dominated social groups such as adventure tourists (Orams) to self-presentation and the behaviours expressed in social research settings (Aquino, Carr, Cheer & Lew, Lacey, Ooi), including discomfort (Lapointe) and safety and security in a conflict-ridden destination (Isaac). Its contributors consider the performance of hegemonic (Carr, Orams, Shepherd), dissonant (Lacey) and heteronormative masculinities (Ooi, Aquino) and examine how performances of masculinity in field research are negotiated and shaped by cross-cultural (Aquino, Cheer & Lew, Isaac, Lacey, Lapointe, McCartney) and situated contexts (Mahler-Coetzee). In addition, it touches on particular issues such as masculinity, identity and mentoring (Cheer & Lew) and fatherhood (Reid, Di Giovine).

These are thought-provoking and important issues, which require consideration at any time and no more so than now. The world is facing multiple crises, notably the global Covid pandemic, climate emergency and the rise of political popularism – in all of which toxic masculinity plays a driving role. There is a long history of revering stereotypical masculinity norms on the political right. Worldwide, conservatives seem to be reacting to the erosion of male privilege and turning to male leaders who embrace masculine stereotypes and are anti-maskers and climate change deniers. That men take greater risk in day-to-day life comes down to harmful gender stereotypes that condition men to glorify strength and fear vulnerability. Men are socialised to be dominant and independent, traits that serve to reinforce patriarchal norms, which present masculinity as a performance. Consequently, men are less likely than women to wear a face mask in the pandemic and more likely to agree that wearing one is a sign of weakness, 'not cool' and 'shameful' (Petter, 2020). At the same time, studies suggest men have bigger carbon footprints on average than women, are more sceptical of climate change, and reject environmentalism because they see it as feminine (Ballew *et al.*, 2018). As our world faces a tipping point, addressing masculinity in all its multiple forms – positive and negative – as this collection does, could not be more crucial, for tourism studies and for every field of enquiry.

Nigel Morgan and Annette Pritchard

References

Aboim, S. (2010) *Plural Masculinities: The Making of the Self in Private Life*. Abingdon: Routledge.

Ballew, M., Marlon, J., Leiserowitz, A. and Maibach, E. (2018) Gender Differences in Public Understanding of Climate Change. *Yale Program on Climate Change Communication Climate Note*. See https://climatecommunication.yale.edu/publications/gender-differences-in-public-understanding-of-climate-change/.

Connell, R. (1995) *Masculinities*. London: Allen & Unwin.

Kimmel, M. (1996) *Manhood in America: A Cultural History*. New York: The Free Press.

Petter, O. (2020) 'Real men don't wear masks': The link between masculinity and face coverings. See https://www.independent.co.uk/life-style/face-masks-men-masculinity-coronavirus-lockdown-boris-johnson-b1077119.html.

Porter, B.A. and Schänzel, H.A. (eds) (2018) *Femininities in the Field: Tourism and Transdisciplinary Research*. Bristol: Channel View Publications.

Pritchard, A. (2014) Gender and feminist perspectives in tourism research. In A. Lew, C.M. Hall and A. Williams (eds) *The Wiley Blackwell Companion to Tourism* (2nd edn, pp. 314–324). Oxford: Wiley Blackwell.

Pritchard, A. and Morgan, N. (2017) Tourism's lost leaders: Analysing gender and performance. *Annals of Tourism Research* 63, 34–47.

Roulston, K. and Shelton, S.A. (2015) Reconceptualizing bias in teaching qualitative research methods. *Qualitative Inquiry* 21 (4), 332–342.

Introduction – Issues in the Field: Masculinities in Masculine Spaces

Brooke A. Porter, Heike A. Schänzel and Joseph M. Cheer

Preface

Soon after finishing *Femininities in the Field*, it became patently obvious that the need for more masculine perspectives on fieldwork was pressing. We were criticised by some who felt that providing a platform for 'men' would weaken the progress in the academy laid by our first (feminine) project. However, our original aims remain unchanged: 'Our call for equality is in recognising that gender issues in fieldwork are deserving of our attention. In doing so, we acknowledge that viewing the field as a masculine space is defunct' (Porter & Schänzel, 2018: 2). In developing this volume, we have intensely explored the complexities and intersections of gender in the field in order to better understand the nuances of performing masculinities within what are regarded as masculine spaces. The decision was made to leverage contributions principally from male researchers (including the enlistment of a male co-editor) and to draw from their collective experiences as researchers actively engaged in scholarly fieldwork. In comparison to the contributions from *Femininities in the Field*, we at times struggled to realise the desired aims that had been established at the outset. Whereas none of the female contributors in our *Femininities* book questioned their femininity, all the men in this book questioned to some extent what it means to be a 'man' today and seemingly distanced themselves from being part of the emergent toxic masculinity. Grappling with one's masculinity was evidently more common than not for contributors to the book, and having to adhere to a narrower definition of masculinity in their academic work appeared to be at odds with most. Some contributors initially wandered into gender critiques of a specific discipline, while others focused on defining, refining or defending masculinity. Consequently, the shifts away from our preconceived expectations to venture into new and expanded spaces for gendered perspectives meant that this volume ended up morphing into a series of granular examinations underlined by reflexive personal perspectives. This contribution to extant knowledge regarding the performance of masculinities in academic research will serve students and scholars well in their exploration of the subjective and objective impacts of gender and gendered performances in fieldwork. In particular, it serves to inform male researchers of the imperative to recognise that the performance of their masculinities in the field requires considered thought, rather than a mere cursory acknowledgement. Moreover, we argue that attending to the overarching implications of how the variegations of maleness manifest in the field can have considerable influence over the success and/or failure of fieldwork endeavours. Furthermore, the focus on maleness provides a space for male researchers to overcome the binding constraints as seen in social constructions and received understandings of what being a man entails, and how this guides and influences processes of academic research.

Masculinity in Masculine Spaces: Tourism Field Research

Masculinity is fundamentally defined as the quality of being male. While this is simplistic, the purpose of this volume is not to challenge the construction and/or the deconstruction of masculinity. Instead our aim is to create a space where the performances of masculinities in academic field spaces can be explored. At a deeper level, Raewyn Connell has long argued that masculinities are multifarious in conception, with untold complexities and contradictions, and that invocations of masculinities as straightforward and simplistic are counterproductive to fostering optimum gender relations (Connell, 1995). The label 'masculine' is commonly accepted as one of two overarching gender categories, with the opposing category defined as feminine. There has been a recent gender revolution placing genders on a spectrum, understanding that a binary approach is inaccurate and restrictive (see *National Geographic: Special Issue on The Shifting Landscape of Gender*, January 2017). Although the gender revolution is seen as a contemporary movement and is contested, gender fluidity and mobility have historical antecedents. Native Americans, depending on the tribes, recognised three to five genders; the gender view was often simplistically translated in English as having 'two spirits' (Brayboy, 2017).[1] Two-spirits people were allowed to move between genders without restriction, and Brayboy (2017) suggests that these non-binary members of Native American communities were highly regarded as possessing traits such as high intellect and compassion, which may have resulted from increased self-reflexivity. In the same vein, the Samoan fa'afāfine embraces the non-binary (Schmidt, 2016) and speaks to Connell's (1995) argumentation that masculinities are complex and multifaceted and far from the singularity that is so often a mark of gendered discourses. The contributions in this volume seek to offer a wide range of perspectives on masculinities. In doing so, while the voices are diverse and at times disjointed, as a collection they contribute to the broader understanding of the disparate ways in which masculinity may manifest in academic spaces.

Gender, in a Western academic context, tends to be thought of as being socially constructed (Criado Perez, 2019; Keller, 1985; Kimmel *et al.*, 2008). Academics have explored variables such as age, sexuality and relationship status and their impacts on gender identifications, gender roles and gender mobility (Golde, 1986; Turnbull, 1986). The development of feminist studies and the continued rise of women's movements has forced us to define, or at the very least explore, what it means to be male. This extends to the conduct of fieldwork and broader aspects of the academy where many spaces remain masculine by default. Still, masculinities, as a field of study, continues to lag behind research on femininities. Some progress is being made, as reflected in the Center for the Study of Men and Masculinities at Stony Brook University, part of the State University of New York system, which began offering a master's degree in 'masculinities studies' in 2019. Masculinity studies is described as offering a framework to break down so-called toxic masculinity, seen as affecting people of all genders. In this book we do not attempt to recast gender and we acknowledge that gender can be experienced as a spectrum. Thus, we have adopted the term 'masculinities' (plural) to acknowledge that there is 'more than one way to be a man' (Bennett, 2015). However, this reflexivity is arguably very much missing in academic debates in tourism research. There are surprising exceptions. For example, Mach's (2019) article on surf tourism included a reflection on how gender impacted the data collection. The overarching question here appears to coalesce around what it means to be male in today's world. We acknowledge that this absence of reflexivity extends to daily life, according to Australian activist for the well-being of families, Steve Biddulph (2013), who argues:

> Most men today live behind masks. They put them on in the morning and keep them on until they fall asleep at night, adopting the clichés of what they perceive a real man to be. The problem is, it's all pretend. (Biddulph, 2013: 1)

Similarly, in his book *The New Manhood*, Biddulph explores how, deep down, men don't know who they are. British artist Grayson Perry in 2017 published *The Descent of Man*, in which he makes a plea for a more expansive blueprint for masculinity, 'one that prizes tolerance, flexibility, plurality, and emotional literacy in the same way that strength, certainty, stoicism have been celebrated in the past'. Today, there is a broader range of role models than ever before but, more importantly, there are a wider number of pathways that should all be considered as equally masculine. For decades now, female writers and theorists have been dismantling their biological gender from the perceived feminine roles that can restrict or harm their lives. But it is only recently that we have started to do this with men too – to see a man as distinct from the concept and construct of masculinity. Perry's and Biddulph's books are a reminder of how awkward popular notions of masculinity can be for modern life and a prompt that a fulsome debate and deep reflexivity for men is much overdue, beyond academia. Agreeing that gender remains misrepresented for many, both male and female, in all spaces, for the purposes of this volume we have focused questions surrounding 'maleness' to performances in academic spaces.

Oakley (2006: 19) opined: 'Theorising *patriarchy* is a minority interest, regarded with mistrust as tainted with the politics of feminism, while the biases in our knowledge due to the politics of *masculinism* go largely unnoticed'. Gender studies have consistently failed to excite the imagination of the media, the general public or the wider social science academy, according to Pritchard *et al.* (2007), because gender is still equated with studies of women's experiences. Pritchard *et al.* (2007) emphasise that tourism academics have an obligation to challenge injustices and inequalities in tourism's material or symbolic domains. There is some momentum in academia with voids being called out. For example, fatherhood as a topic of research is more or less absent in tourism studies (Schänzel & Smith, 2011) with little forthcoming. Yet we know when looking at the contributions in this volume that the gender debate needs to be broadened to include all genders. But this debate is not solely the concern of women because enduring patriarchies and emergent toxic masculinities constrain potentialities for those receiving them and those *performing* them. A way forward is to bring a kind of balance (in ourselves, our communities and our world). Rather than replacing one 'archy' (patriarchy) with another (matriarchy), instead we aim to embrace a more holistic thinking: 'it is about reuniting the head, heart and body' (Pritchard *et al.*, 2007: 9). Increasing self-reflexivity, regardless of gender, is needed, noting that studies of masculinities and its accompanying roles (e.g. fatherhood) are also overdue. In the development of this volume we encountered difficulties in recruiting male contributors who were prepared to reflect on their fatherhood roles in field research. A similar sentiment was mirrored in Marotte *et al.*'s (2011) collection of essays on fatherhood in academia. They noted contributors withdrawing submissions and or declining submissions due to reluctance to self-reflect and discuss their feelings openly.

Although there is a freedom associated with autoethnography, and many of the authors in this book have leveraged that either intentionally or unintentionally (see also Porter & Schänzel, 2018), the process may not be comfortable and could evoke issues of privilege (see, again, Marotte *et al.*, 2011). Noting that some scholars have criticised such unconventional approaches as unscientific or less rigorous, we argue that the field is a far from sterile environment and is a space where researcher and participants are in constant negotiation. In their overview of the processes and product of autoethnography, Ellis *et al.* (2011: 274) define autoethnography as 'one of the approaches that acknowledges and accommodates subjectivity, emotionality, and the researcher's influence on research, rather than hiding from these matters or assuming they don't exist'. Hence we argue that the exploration and discussion of these personal subjectivities contribute to the transparency and trustworthiness of the fieldwork. Gender and gendered performances, as well as other individualities (e.g. identities, positionalities, subjectivities) of the researcher, are generally discernible to participants. Thus, the common omission of these variables from the literature creates fertile ground for discussion. It seems men are less comfortable with reflexivity. However, this discomfort may simply be a default to performing masculinities in masculine spaces. For if gender and space are aligned, one could argue that there is less space for dissonance. This interpretation is supported by the contributions found in Part 2 on 'Performing Heteronormative Masculinities'.

The idea of a gendered privilege accompanies masculinity in a masculine space. Caroline Criado Perez (2019), in *Invisible Women: Exposing Data Bias in a World Designed for Men*, states that there are three themes that crop up again and again: the female body; women's unpaid care burden; and male violence against women. These are issues of such significance that they touch on nearly every area of women's lives, including conducting research in the field and working as an academic. This is confirmed by our findings in *Femininities in the Field* (Porter & Schänzel, 2018), with themes of conduct and attire, sexual harassment, personal safety and accompanied research with children. But the same is not true for men because men do not have female bodies or, often, the same care responsibilities. And while men may have to contend with harassment, this violence typically manifests in quite a different way from the violence faced by women. Overall, these differences in fieldwork experiences have been largely ignored, and academia has proceeded as if the male body and its attendant life experience are the gender-neutral default. It is only through more reflective exercises that the gendered nature of fieldwork and research is beginning to emerge.

There is, however, a word of caution when widening the discussion of masculinities and fieldwork, as highlighted by Vanderbeck (2005), who reviewed 'the ways in which male social scientists have tended to write about their fieldwork experiences, emphasising how many ostensibly reflexive accounts actually serve to reinforce the author's own hegemonic masculine position' (Vanderbeck, 2005: 388). A risk in creating a space for masculine gendered experiences is the danger in further validating hegemonic ideals and conformity while asking others to expose their vulnerabilities in the ways in which they might fail to do their gender 'right'. Following the style and approach of the precursor, *Femininities in the Field*, our intent is that the reflexivities articulated should bring acutely sensitive issues to light, and to enable this, the call is for a culture of trust and gender maturity within which such openness can flourish. This too, may come with risk. What if our tourism academia is not ready for men to bear their souls? This is highlighted by Brene Brown's (2012) work about acknowledging male vulnerabilities, where the number one shame trigger for men is to be perceived as weak, with consensus from her male participants that families do not want to see their men as vulnerable and real. What if society is not ready for male vulnerabilities?

The overarching aim of this book is to advance the attention of gender-entwined biases in fieldwork, this time through a masculine lens. The contributions in this volume are self-defined, self-reflexive, individualised first-person accounts describing the constructs of varying masculinities and the sometimes fluid and contextual performance of masculinities. These reflexive accounts of gendered field experiences further the call for gender acknowledgement in research and compile descriptions that will aid tourism researchers and other transdisciplinary scholars.

Intentions and Structure of the Book

The chapters that follow span various geographical locations in Europe, Asia, Australia, Africa, Oceania and North America. The contributions are based on experiential and reflexive analyses of the authors' previous field experiences and span a wide background with reflections on masculinities coming in many personal forms. The cases presented in this volume are organised into four parts:

Part 1: Hegemonic and Heteronormative Masculinities
Part 2: Performing Heteronormative Masculinities
Part 3: Situated Masculinities
Part 4: Paternal Masculinities

The first part, 'Hegemonic and Heteronormative Masculinities', looks at what it means to be a (manly) man. It begins with Jack Shepherd's call for emotional writing from male tourism scholars. He traces his lived experiences of performing field research in Palestine and Israel, and tells us about a conference presentation where he questioned his research approach. Shepherd's experiences are followed by Neil Carr, who trifles with his (possibly hegemonic) male-ness. In doing so, he uncovers privileges, such as access to taboo research topics, along with barriers such as difficulties in accessing child participants as a male. In the next chapter, Mark Orams reflects on his learnings about masculinity in New Zealand and how these have influenced his identity as a participant in nature-based adventure recreation – his current research focus. He explores how, ultimately, his identity and masculinity impact his ability to approach these otherwise difficult-to-access participant groups. Looking at transitions in masculinities, Jacques Mahler-Coetzee 'meditates' on masculinities in varying temporal and contextual experiences, some of which have impacted his access to participants.

In the second part, 'Performing Heteronormative Masculinities', contributors reflect on gender and sexuality and their influence on fieldwork and gendered 'performance'. Richard Aquino unpacks the colonial history of the Philippines (his home nation and field site) and how the resulting patriarchal social ideas influence his performance of masculinities in the field. In the chapter that follows, Can-Seng Ooi revisits awkward field encounters in Asia in which he 'tacitly ignored and marginalised' his masculinity, in some cases attempting to take on an androgynous persona.

Part 3, 'Situated Masculinities', explores the intersections of gender, ethnicities and cross-cultural experiences. Joseph Cheer and Alan Lew converse on how gendered and mixed ethnicities have afforded them mobility as well as how temporal, structural and conceptual constraints have influenced the performative aspects of masculinities. Dominic Lapointe grapples with conflicting social and ethnic dynamics in his negotiations of self and how these interplay with his participants. Patrick McCartney describes the challenges of accessing yoga participants in a foreign Japanese culture and navigating a space that is culturally feminine. Gary Lacey reflects on his experiences as a cross-cultural and mixed-raced researcher in Kenya, and the resulting variable dependencies of masculinities in dissonant contexts. Rami Isaac delves into the implications of fieldwork in occupied Palestine and the associated personal impacts.

In Part 4, 'Paternal Masculinities', Stuart Reid reflects on accompanied fieldwork in Sweden as a father-researcher and sometimes head of household and what it meant for his fieldwork relocations. Michael Di Giovine explores the intricacies of fatherhood, culture and identity in an ethnographic entry to fatherhood.

What remains is that the emergence of masculinities as a critical issue for consideration in the construction of research planning and applied fieldwork approaches underlines this volume. This is exemplified by contributors to the volume who, up until authoring their respective chapters, had not considered the performance of masculinities as critical to their work. This alludes to an academic environment which has largely failed to embrace the diverse privileges and strictures that have shaped and continue to shape male researchers and their field practice. This volume showcases why the wider academy must embrace 'masculinities training', as it were, for whether male researchers are acquainting themselves sufficiently with their participants and eliciting optimal responses is brought into question. Indeed, in being better acquainted with their masculine selves, reflexivity, empathy and compassion may just be heightened, generating more fertile interactions with participants across all genders.

In many respects, one of the inadvertent outcomes of the continued fight for gender equality has been a shaking up of gender relations and a resetting of the power pendulum, bringing to light the anguished and frustrated discourses of feminist scholars appealing for justice and equality. That said, this has tended to lead to oversimplifications about gender and pitting genders in opposition on account of disproportions in power, privilege and access. For many of the contributors to this volume, their encounter with masculinities within the formal scope of an authored chapter was unprecedented; up until now, such discussions have been largely muted and under-acknowledged.

As Can-Seng Ooi muses in his chapter, masculinity and what it means to be a man have been reflected in many studies, yet scrutinising how this is integrated into fieldwork planning and the application of field research has been given little attention. This reflects how masculinities and their abiding influence on the work that male researchers do has only been given prominence outside the technical aspects of research. This can be seen in the absence of its mention in discussions about methods and wider methodological considerations. This, however, is not so for considerations around femininities and the structures under which female researchers have toiled. Certainly, the time has come for masculinities to be seen as just as critical as femininities are in academic practice, in that they offer male researchers markers that might lead to more fruitful and mutually beneficial gendered exchanges.

Note

(1) According to the description of Brayboy (2017), which is based on oral histories: 'Native Americans traditionally assign no moral gradient to love or sexuality; a person was judged for their contributions to their tribe and for their character. It was also a custom for parents to not interfere with nature and so among some tribes, children wore gender-neutral clothes until they reached an age where they decided for themselves which path they would walk, and the appropriate ceremonies followed.'

References

Bennett, J. (2015) A master's degree in ... masculinity? *The New York Times*, 8 August. See https://www.nytimes.com/2015/08/09/fashion/masculinities-studies-stonybrook-michael-kimmel.html.

Biddulph, S. (2013) *The New Manhood: The 20th Anniversary Edition*. Sydney: Simon & Schuster Australia.

Brayboy, D. (2017) Two spirits, one heart, five genders. *Indian Country Today*, 7 September. See https://newsmaven.io/indiancountrytoday/archive/two-spirits-one-heart-five-genders-9UH_xnbfVEWQHWkjNn0rQQ/.

Brown, B. (2012) *Daring Greatly: How the Courage to Be Vulnerable Transforms the Way We Live, Love, Parent, and Lead*. New York: Penguin.

Connell, R.W. (1995) *Masculinities*. Berkeley, CA: University of California Press.

Criado Perez, C. (2019) *Invisible Women: Exposing Data Bias in a World Designed for Men*. London: Chatto & Windus.

Ellis, C., Adams, T.E. and Bochner, A.P. (2011) Autoethnography: An overview. *Historical Social Research* 36 (4), 273–290.

Golde, P. (1986) Odyssey of encounter. In P. Golde (ed.) *Women in the Field: Anthropological Experiences* (2nd edn) (pp. 67–96). Los Angeles, CA: University of California Press.

Keller, E.F. (1985) *Reflections on Gender and Science*. New Haven, CT: Yale University Press.

Kimmel, M.S., Aronson, A. and Kaler, A. (eds) (2008) *The Gendered Society Reader*. New York: Oxford University Press.

Mach, L. (2019) Surf-for-development: An exploration of program recipient perspectives in Lobitos, Peru. *Journal of Sport and Social Issues* 43 (6), 438–461.

Marotte, M.R., Reynolds, P. and Savarese, R.J. (eds) (2011) *Papa, PhD: Essays on Fatherhood by Men in the Academy*. New Brunswick, NJ: Rutgers University Press.

Oakley, A. (2006) Feminism isn't ready to be swept under the carpet. *The Times Higher Education Supplement*, 3 March, pp. 18–19.

Perry, G. (2017) *The Descent of Man*. New York: Penguin.

Porter, B.A. and Schänzel, H.A. (eds) (2018) *Femininities in the Field: Tourism and Transdisciplinary Research*. Bristol: Channel View Publications.

Pritchard, A., Morgan, N., Ateljevic, I. and Harris, C. (2007) Editors' introduction: Tourism, gender, embodiment and experience. In A. Pritchard, N. Morgan, I. Ateljevic and C. Harris (eds) *Tourism and Gender: Embodiment, Sensuality and Experience* (pp. 1–12). Wallingford: CABI.

Schänzel, H.A. and Smith, K.A. (2011) The absence of fatherhood: Achieving true gender scholarship in family tourism research. *Annals of Leisure Research* 14, 143–154.

Schmidt, J. (2016) Being 'like a woman': Fa'afāfine and Samoan masculinity. *Asia Pacific Journal of Anthropology* 17 (3–4), 287–304.

Turnbull, C.M. (1986) Sex and gender: The role of subjectivity in field research. In T.L. Whitehead and M.E. Conaway (eds) *Self, Sex, and Gender in Cross-cultural Fieldwork* (pp. 17–27). Urbana, IL: University of Illinois Press.

Vanderbeck, R.M. (2005) Masculinities and fieldwork: Widening the discussion. *Gender, Place & Culture* 12 (4), 387–402.

Part 1
Hegemonic and Heteronormative Masculinities

1 It's Okay to Cry: Encouraging Emotional Writing Among Male Tourism Scholars

Jack Shepherd

> What are we saying to a boy when we tell him to 'man up' or 'act like a man'? At its most benign, we might be saying: do the thing that needs doing even if you don't want to do it. But more often, when we tell a boy to 'act like a man', we're effectively saying 'Stop expressing those feelings'. And if the boy hears that often enough, it actually starts to sound uncannily like 'Stop *feeling* those feelings'. It sounds like this: 'Pain, guilt, grief, fear, anxiety: these are not appropriate emotions for a boy because they will be unacceptable emotions for a man. The skills you need to be your own emotional detective – being able to name a feeling and work out why you're feeling it – you don't need to develop those skills. You won't need them.'
>
> Robert Webb, 2017: 37[1]

Boys Will Be Boys

'Stop crying', he wrote. I stared at my phone, feeling the blood boil up into my ears, making me feel hot and flushed. I am 26. Another man my age has just tried to shut me down in a conversation with a classic 'masculine' rebuke. I was not crying. I was explaining why I felt he was wrong, and how his accusations made me feel. For some reason, I let the phrase get to me. I found my brain saying over and over, 'Well I am not crying!', as if trying to prove (to him and myself) that I was still 'Acting Like A Man' or should 'Man Up', and that I should not care that this conversation was bothering me. Why is it that I care? Why do I feel pressure not to talk about emotions?

Whether I like it or not, I have always been someone who feels a lot – from falling head over heels in love to day-crippling anxiety. As a boy, I felt greater affinity with the girls at school who would rather play 'mummies and daddies' than try to kill each other in the 'adventure playground'.

14 Part 1: Hegemonic and Heteronormative Masculinities

Figure 1.1 Is this where our journey to the 'masculine social science' begins? My first day at school, Eastbourne, 1996 (from the author's private collection)

Of course I also enjoyed a good 'rough and tumble' with the other boys, up to the point when at 14 years old it became known as 'Fight Club', which was more a macabre set-up than honest hurly-burly. That willingness to hang out with the 'fairer' sex, to excel at gymnastics and to love drama (in particular, involving myself in the school musicals) landed me with a number of nicknames, most commonly 'gay lord' – which, although not true (as if such a thing exists!), stuck around for quite some time. *Boys will be boys*,[2] and so it continued. To be a 'gay lord' clearly meant to be 'feminine', which for them meant being emotional and enjoying the company of girls – not something a self-respecting boy would (or should) want to do.[3]

From boyhood, we quickly learn how we should perform masculinities and that masculinity is, in essence, simply the opposite of what is feminine (Aboim, 2010). Unlike sex, which is a biological categorisation, gender is a form of practice – a structure of socially and culturally based beliefs and practices that create differences between men and women (Aboim, 2010; Figueroa-Domecq *et al.*, 2015; Munar, 2018). Through a process called gender socialisation – whereby 'individuals develop, refine and learn to "do" gender through internalising gender norms and roles as they interact with key agents of socialisation, such as their family, social networks or social institutions' (Balvin, 2017) – the lines between gender and sex become blurred, creating a situation where biological men should adhere to what is considered 'masculine' and biological women to what is 'feminine'. Boys, for example, are not born with an innate desire to love football and look muscle-bound any more than girls are; it is something they are taught over time through gender socialisation.[4] Gender socialisation is omnipresent and varies according to the society you live in. In the West,[5] boys should wear blue and girls pink[6]; boys should play with Action

Man™, girls with Barbies™; boys should look after their cars, girls their dolls.⁷ The all-pervasive assigning of gender roles and characteristics, drilled into us from the windows and aisles of the shops we visit, to the plots and casting of the films we watch (any classic Disney film will give you a clear picture …) leaves society with little room for questioning what it is to be masculine and feminine – they are so unescapable as to be unquestionable. Therefore, boys in the West grow up in a society that teaches them that they should look like a Ken doll™, fight like Action Man™ and have the emotional depth of Prince Charming.⁸

Learning to Be a Man

As boys grow up, so does their masculinity. It becomes stricter, its expectations heavier. It is okay to bumble your way through masculinity as a boy, but not as a man. As a man, you have to start perfecting the art of what it means to be masculine. As Robert Webb superbly puts it, 'It is not so much Act Your Age but Act Your Gender' (Webb, 2017: 121). It is for this reason that I gave up gymnastics at 14 – my secondary school hardly encouraged it and, well, a teenage boy in a leotard is hardly tolerable. It is for this reason that at 15 I had to react in complete horror when I discovered that my mother had asked the school to intervene despite my crying on the way home many times from the bullying. It is for this reason that any disagreement at school was always negotiated with a fight (albeit 'play fight'). Conversely, as a teenager, I learnt some key lessons about masculinity: don't show any weakness (physical or mental); don't show emotions (they will be used against you); and don't, EVER, get help if you need it (because taking it on the chin is part of Acting Like A Man).

The supremacy of obstinacy and the rejection of emotion as key cornerstones of 'Western' masculinity entrench themselves in manhood. It is no coincidence that men are three times more likely than women to commit suicide in the UK (Samaritans, 2019), that men are a third less likely to seek primary care consultation (Wang *et al.*, 2013) or that the more 'masculine' the man, the less he would report symptoms to a doctor (Himmelstein & Sanchez, 2016). Men struggle to speak about the things that trouble them or affect them, because since childhood they have been discouraged from doing so. To summarise, Webb writes:

> Men in trouble are often in trouble precisely because they are trying to Get a Grip and Act Like a Man. We are at risk of suicide because the alternative is to ask for help, something we have been repeatedly told is unmanly. We are in prison because the traditional breadwinning expectations of manhood can't be met, or the pressure to conform is too great, or the option of violence has been frowned upon but implicitly sanctioned since we were children. We are dependent on booze when we try to tilt the table, try to change the chemistry in a way that is harmful, counterproductive and, of course, widely accepted as tough and manly,

irrespective of whether the impulse comes from conformity or rebellion, from John Wayne or James Dean. We die younger than women because, for one thing, we don't go to the doctor. We don't take ourselves too seriously. We don't want to be thought self-indulgent. The mark of a real man is being able to tolerate a chest infection for three months before laying off the smokes or asking for medicine. (Webb, 2017: 236)

Not sharing your emotions can be lethal in one's personal life, so why then do we discourage the exploration of emotions within our professional life in tourism academia?

We Are All Part of 'The Trick'

Being emotionally introspective and being masculine often appear to be mutually exclusive. And if we, as tourism scholars, assume that we are somehow outside the masculinities trap, what Webb calls 'The Trick',[9] then we are being naïve at best and darn right foolish at worst. Gender conditioning happens to us all, whether we decide to be conscious of it or not. The recent publication, *Femininities in the Field* (Porter & Schänzel, 2018), was a timely intervention in our understanding of how gender is performed and perceived when conducting tourism research; of course, it took us blokes another couple of years to get round to looking at how *our* gender affects the way we do research. It is my argument in this chapter that there is a clear link between the masculinity spoken about in Webb's *How Not To be a Boy* (2017) or Grayburn's *Boys Don't Cry* (2017) – a masculinity that teaches boys and men not to talk about emotions – and the 'masculine' approach to tourism studies (and social science more generally), which is so often derided by scholars as positivistic, objective and emotionless (Buda & McIntosh, 2012; Pritchard & Morgan, 2000; Pritchard *et al.*, 2011; Thien, 2005; Wall, 2006). Scholars who criticise a 'masculine' approach often fail to state explicitly why the approach they are deriding is 'masculine', which tells us rather depressingly that they consider it a given that the reader assumes a masculine approach equals detached, emotionless and authoritative. Conversely, I therefore argue that the emotional constipation caused by 'The Trick', which in its worst forms leads to depression and suicide, is the same emotional constipation that is affecting the way we produce and write about knowledge in tourism.

The reason why it is so important that we are aware of this link is because, funnily enough, tourism studies is still dominated by men. Numerically, men dominate – the editorial boards of tourism journals are predominantly comprised of men (Figueroa-Domecq *et al.*, 2015; Pritchard & Morgan, 2007), the number of keynote speakers (globally speaking) continue to be mostly men and, as Ek and Larson (2017) demonstrated, the authors held up as the 'greatest of the great' of tourism studies in the journal *Anatolia* were overwhelmingly male. Culturally, men dominate too. As Ek and Larsson (2017) suggest, the tourism academy seems to act more like the ancient Greek *oikos* (household) than the *agora* (forum), where the masters are to be respected and their works referenced and replicated, with tradition dominating our methods of work. Feminist methodologies, which have argued for a more inclusive, emotional and democratic social science, have largely been left by the wayside (Munar, 2018; Wilson & Hollinshead, 2015). Therefore, given that men continue to hold the positions as gatekeepers of tourism knowledge, it is surprising – shocking even – that we have not mused on how being (masculine) men affects the way we write about tourism.

The Consequences of 'The Trick' on the Way We Produce Knowledge

So if we have learnt that masculinities discourage emotion and that men are the gatekeepers of tourism knowledge, it is no surprise to see that tourism studies shows a chronic disregard for emotional writing (Buda, 2015; Thurnell-Read, 2011). Some have attributed this emotionless body of research to the fact that tourism has often been, and continues to be, housed within business and economics departments that constrain the more human side of tourism (Beeton, 2016). Others lay the blame at the feet of a Western epistemology that has viewed the placing of emotion and reason within social science as incompatible (Davenport & Hall, 2011; Emerland & Carpenter, 2015; Gilbert, 2000). Another – complementary – way of looking at it would be to see that men are just generally not comfortable talking about emotions due to the gender socialisation we discussed earlier, and in an academy dominated by men it is not surprising to see a lack of emotional writing. It is for this reason that we see statements such as 'Even to speak of hope and love as academics however makes us vulnerable as this is associated with weakness, irrationality and emotion' (Pritchard *et al.*, 2011: 951).

This is a tragedy in many ways – firstly, and most importantly, because emotions are integral to our understanding of the world. Everything we understand of the world is mediated through emotions which 'lie at the intersection of the person and society' (Denzin, 1984: 61; Thien, 2005). We experience our research in the field not just intellectually but also emotionally (Gilbert, 2000), and by leaving emotions out of our writing we are not only denying the humanity within all of us, but we are also actually abandoning half of the data the field has provided to us. Crucially, tourism is an emotional experience. Tourism is about far more than simple supply and demand, smart destinations and marketing – it is about the taste of that pizza in Naples, the feel of the rough stones of Pompeii, the first sip of a cold beer after a long hike in the summer. Denying the emotional side of tourism negates an exploration of the most important aspects of tourism: the embodied experiences – good and bad – that make tourism memorable (Pritchard *et al.*, 2011; Thurnell-Read, 2011).

The second great tragedy that results from this denial of emotion is the production of texts that are stale, author-absent and lifeless (Pritchard *et al.*, 2011). It is amazing that despite being called a 'social science', many texts within tourism studies read as positively anti-social – denied the opportunity to socially/emotionally engage due to a restrictive blueprint that goes somewhat like this: Introduction – Literature review – Methods – Results – Discussion – Conclusion – Thanks for not falling asleep. Readers have been left disenchanted (Burlingame, 2019), struggling to find an authentic voice in the text that gives us a sense of place and person. It was until very recently even that authors in major journals within tourism were forbidden to use the pronoun 'I'. What a bizarre situation – that the writer had to actively write themselves out of the text! The reason for such absurdity comes from another consequence of the primacy of masculinity in our academy – 'the façade of objectivity' (Wall, 2006: 147). Much as how men might like to pretend to not have 'issues' or emotions to talk about, objectivity pretends that the researcher can enter the field as some kind of emotionless, detached observer – creepily lurking in the shadows, extracting knowledge while remaining apart from the social world of the researched (Pritchard *et al.*, 2011). This stance not only ignores the basic fact that all our knowledge comes from a subjective understanding of the world (Wright, 2018), but it is also crucially creates multiple layers of distancing: between researcher and researched, the reader and researcher and, ergo, the reader and the researched (Doty, 2010; Miraftab, 2004) – leading us to distanced texts that do not capture the essence of what we want to describe.

The third issue with rejecting emotions in our research is that we dehumanise those we study. By not permitting emotions in our work, we not only remove our own individuality but also the individuality of those we study. We often explain tourism in terms of master narratives, boxing people up into categories such as 'mass tourist', 'dark tourist', 'ecotourist', 'danger-zoner' and 'solidarity tourist', which tend to overshadow the multi-experiential experience of tourism (Noy, 2007; Shepherd *et al.*, 2020). Master narratives are helpful little devils when all you want is simplification, but the boxing up of people into categories or typologies is highly problematic. As Grant (2018, personal correspondence) explains:

> Conventional, typologising qualitative research does violence to people's experiences through cultural colonisation (imposing typologies), cultural appropriation (using people's experiences as the basis for typologising), and cultural violence (ignoring/marginalising the richness of people's experiences in the act of typologising. (Grant, 2018, personal correspondence)

As our academy tries to move away from the 'bean-counting' research of working out who tourists are and what they do (Botterill, 2003: 99), and starts to take a more critical approach to the socioeconomic phenomena of tourism, the time is nigh when we drop the boxing up of people's lived experiences and start embracing the ephemeral, context-specific and subjective nature of the tourist experience.

How Do We Rescue Tourism from 'The Trick'?

How do we bring emotions back into our scholarship? How do we do justice to the fact that emotionality is central to our research process (Denzin, 1984; Gilbert, 2000)? How do we heed the call of our sibling publication, *Femininities in the Field*, and 'bring our humanity and its emotions and sensibilities closer to the centre of what we do as researchers' (Porter & Schänzel, 2018: xi)? My proposition is that autoethnography is one tool that could help us to recover from some of the damaging effects The Trick has had on our academy, given that it puts reflexivity and emotions central to our understanding of culture.

In *Boys Don't Cry* (2017), Tim Grayburn decides to overcome the suffocating pressures masculinity placed on him not to talk about his emotions (more precisely, his depression anxiety disorder) by starting a theatre play with his girlfriend Bryony called *Fake It 'til you Make It*. The play, the product of a few months' work in their living room, became a huge success at festivals in Australia and the UK, prompting hundreds of men to write to Tim about their difficulties with depression and how his story helped them to better understand their own lives. Tim's show connected with people because stories have the power to connect people. Stories are, if we are to follow Yuval Harari's (2014) worldview, the basis of the success of our species.[10] Autoethnography is a method that uses our personal stories to explore societal issues (Ellis & Bochner, 2006). It is 'an approach to research and writing that seeks to describe and systematically analyse (graphy) personal experience (auto) in order to understand cultural experience (ethno)' (Ellis *et al.*, 2011: §1). It begs the question, 'How does my own experience of my own culture offer insights into my culture, situation, event and/or way of life?' (Patton, 2004: 48). This results in highly reflexive and often emotionally charged work, where we explore not just what we think but why we think it and how – not something masculinities has helped us to do (Sparkes, 2002).

How does autoethnography help scholars overcome some of the issues we have associated with masculinity? First of all, autoethnography – which emerged out of the Triple Crisis (Denzin & Lincoln, 2000), one component of which was the crisis of representation – does not claim to speak for others or impose its narrative on others. There is no cultural colonisation, no typologising of lived experience, no 'mansplaining' of another's biographical data. This said, just because autoethnographic texts only speak from one point of view does not mean they are only relatable to one being. All individual voices are created within 'a societal framework of co-constructed meaning' (Wall, 2006: 155), and so good autoethnography seeks to get the reader to enter the world of the writer through engaging storytelling that makes the cultural aspects of personal experience come alive. Critics who call autoethnography narcissistic or boring (Delamont, 2007) seem to forget the huge amount of knowledge we gain from first-person accounts, whether that it is from novels, historical sources or religious texts.

Autoethnography also encourages scholars, particularly men, to frustrate the work–home binary that conventional masculinity and the academy has placed upon them (Reinharz, 1997).[11] A male researcher in the field is far more than just a tourism researcher: he is perhaps a father, a husband, a brother, a son, an activist, a sufferer of mental or physical illness, and of course, often a tourist! Autoethnography criticises the idea of a stable, singular, authorial voice in a text and encourages us to embrace and reveal the multiple selves that we bring with us into the field, allowing the 'cacophony of voices' (Chaudry, 2009: 139) that make up the self to speak, not just the 'accepted' voice that the journal wants to hear (Burlingame, 2019). This results in a far more genuine, situated knowledge being produced.

Lastly, autoethnography gives scholars the platform to express their authentic selves, which is usually denied to them because of the formatting of social science publishing. In a world of out-of-context quoting and tight word limits, autoethnography, inspired by the literary and narrative turns in social science, encourages us to take the time to develop a story through thick description using narrative techniques. It allows the voice(s) in the text to breathe organically in the medium best suited to that voice – mediums that range from poetry to performance. Academics are storytellers, yet they are generally not encouraged to tell that story in an accessible, engaging or creative way, with interpretation seen as more important than communication (Burlingame, 2019; Ellis *et al.*, 2011; Shondell Miller, 2008). In autoethnographic thick description, the author can create affect through 'heartful' (Sparkes, 2002) accounts whereby the reader can feel empathy, or is roused by the honest, often confessional, accounts of the writer.

Male Scholars and Autoethnography in Tourism

Even though autoethnography might contribute towards chipping away at the 'masculine' tourism academy that we have, autoethnography can be a difficult pill to swallow for men who have grown up in an academic culture of dominant masculinity[12] (Campbell, 2017; Ellis & Bochner, 2006). WHAT?! EMOTIONS?! SUBJECTIVITY?! VULNERABILITY?! NOT ON MY WATCH! Traditionally, we seem to have been more than happy to take and interpret the stories of others yet shy away from what our own stories, experiences and emotions can tell us about the culture we study. Yes, gentlemen, that would involve opening up the closet, perhaps moving the skeletons about, but the result would be that men can prove they too can be 'people orientated', 'reflexive' and good readers of people – most importantly ourselves! – qualities *Femininities in the Field* suggest are the preserve of women (Porter & Schänzel, 2018: 5).

So far, autoethnography within tourism is still in its infancy, yet the corpus of authoethnographies relating to tourism is growing as the years go by. In order to see the lay of the autoethnographic land, and to test a hypothesis that men would shy away from autoethnography, I conducted a bibliometric study of authoethnographies published in tourism-related journals and books, using the peer-reviewed literature databases Elsevier, JSTOR, Web of Science, Scopus and ProQuest Social Sciences. Using the keywords 'tourism' or 'tourist' coupled with 'autoethnography' or 'autoethnography', I managed to find 57 peer-reviewed journal articles or book chapters that state that autoethnography was the method used. To my surprise, I learnt that men have not in fact been shying away from using autoethnography within tourism, as they are listed as authors on 38 of the 57 works found. They appear as solo authors on 19 of the 57 works and co-authors on 17. Women solo author the remaining 21 works. This said, given the importance of the self in authoethnographies, I felt it was important to dig deeper into the co-authored papers to see who was contributing the first-person data to the study. When I did this, it became clear that men were generally taking the back seat, with women providing the autoethnographic text and men usually analysing 'the data'. Of the 17 co-authored papers, only four times are men the first author, and in all the other 13 papers bar two, men do not contribute any autoethnographic data to the study. This suggests a slight male coyness in that they are willing to feature on papers dealing with autoethnography yet are not willing in many cases to take the reflexive driving seat.

It is far more interesting, however, from my point of view, to look at what types of authoethnographies men are producing. Autoethnography calls on the huddled masses of academia yearning to breathe (epistemologically) free, to lose their (stylistic) chains, and yet the call remains faintly heard. The majority of male authoethnographies in tourism tend to abuse the inherent flexibility of autoethnography by bending the method towards more conventional social science approaches. Perhaps this stems from a desire to be on familiar ground or from the emotional constipation caused by The Trick. Either way, the result is a number of 'authoethnographies' where personal stories are broken down and lost in a mire of theoretical interruptions, and often coded and analysed in the same way as interview or focus group data would be. This style of 'analytic autoethnography' (Anderson, 2006) shies away from emotions and evocative writing, to the point that it no longer becomes clear whether we can consider it an autoethnography at all (Ellis & Bochner, 2006)! Furthermore, scholars often use autoethnographic titbits combined with other data forms such as interview testimonies, possibly thereby hoping to ward off accusations of self-indulgence by providing a more 'balanced' narrative. Most depressingly, some authors have even struggled to overcome the objective diktat of not using the pronoun 'I', which is quite astonishing given the importance placed on expressing the self in autoethnography (Ellis & Bochner, 2006). There are, however, notable exceptions. Chaim Noy's (2007) poem of a family trip to the Israeli Red Sea coast introduced us to how narrative techniques could be used to understand the multi-experiential nature of tourism. DeMond Shondell Miller's (2008) account of a return to his beloved Louisiana following the devastation of Hurricane Katrina got to the core of what performing tourism means in sites of death and destruction, in much the same way as Anandam Kavoori's (2018) text did at the Cheung Eook Killing Fields in Cambodia. Although the majority of male scholars using autoethnography engage it with extreme trepidation, dipping their toes into the perceived 'naughty pool' of subjectivity, others have felt a sense of liberation (Botterill, 2003) – authors such as Gary Best (2017: 63), whose autoethnographic journey into automobile tourism led him to remark: 'The autoethnographic "I" initially unsettled me after such a long commitment to research objectivity but once underway, I had a sense of shaking off the shackles; free at last.' It triggered a sense not just of freedom but also of surprise at the vividness and thought-provoking nature of the data that autoethnography was able to accrue as compared to conventional qualitative methods (Waterton *et al.*, 2018).

Scraping the Emotional Barrel in Palestine

My own journey with autoethnography began during a research trip to Palestine, not that long ago. I was staying at The Walled Off Hotel in Bethlehem, a hotel created by British street artist Banksy, which intends to inform tourists about the Israeli Occupation of the West Bank and, importantly, to provoke a reaction through art. The more time I spent at the hotel, the clearer it became that I was not simply a researcher looking at how alternative tourism spaces could challenge master narratives of conflict. I was more. I was inescapably a tourist in a conflict zone – and also a Brit. I was the perfect target audience for this hotel.[13]

As I sat in the lobby of the hotel, conversing with other tourists about the hotel and their experience of tourism in the contested town of Bethlehem, I realised quite quickly that obtaining 'honest' data from visitors would be tricky. Often I felt they told me what they thought an acceptable statement would be or something that a researcher, an activist or the hotel would want them to think. I felt I was not getting the full picture of their experiences at the hotel, especially their emotions which they might perceive as inappropriate or too personal. It was out of this slight frustration that I thought back to a recent course I attended on autoethnography. I wondered: could the same method that had given our

Figure 1.2 Positionality pains. Bethlehem, February 2018 (taken by the author)

lecturer Alec Grant (2010) the freedom to explore his battles with alcoholism be used to explore the emotional geographies of a commodified conflict zone? I gave it a shot.[14] In doing so, autoethnography gave me immense freedom to come to terms with the hundreds of questions circling in my head, questions coming from my time in Bethlehem, informed by many years' experience as an independent tourist and from researching tourism in contested spaces. I decided to use 'self as instrument' to explore the nexus of theory and emotionally lived experience.

Putting Emotions into Motion: A Trip from Bethlehem to Tel Aviv

What makes tourism so interesting within the Israeli-Palestinian context is the fact that tourists have immense mobility capital in a small geographical area that is bitterly divided. Tourists can simply 'pop over' to the West Bank for a day and be back in Tel Aviv for supper. This is an incredible power in a region dominated by spatial complexity,[15] spatial exclusion and sharp religious, cultural and national borders. The mobility capital (Kaufmann *et al.*, 2004; Sheller, 2014) that tourists have puts them in a great position of power vis-à-vis local people. They can weave in between worlds in the space of minutes; they can hear two sides of the same story in divided cities such as Hebron or Jerusalem[16] and, with just their presence, influence local security situations.[17] Yet in an effort to demonstrate how this power could be utilised by one or the other competing powers in the Israeli-Palestinian context, scholars have largely ignored the experience of the tourist with regard to this privilege. In the following example of a short autoethnographic account of a trip from Bethlehem, Palestine to Tel Aviv, Israel, I attempt to highlight the power and poetics of being a tourist capable of moving between worlds in a region dominated by borders of all kinds.

Three coffees

9am: I order an Americano at a café by the Church of the Nativity and take one of the green plastic seats on the balcony overlooking the landscape surrounding Bethlehem. The wind has a chill to it and I brace against the cold, pulling the ends of my leather jacket over my hands. As I wait for my coffee, I survey the scene. Sprawling Palestinian suburbs. Bypass roads. The wall. The fortress-like Jewish settlements crowning the hilltops. The canvas of the occupation. The dystopian view has an allure of sorts and I become lost in thought: what happened to the biblical landscape of my childhood? What would Jesus make of the glitzy church next door? What am I doing on Monday back home? Suddenly, I am interrupted by the arrival of the coffee – Italian style, not too much added water, just how I like it. Nice coffee, terrible occupation.

11am: The taxi drops me in front of Checkpoint 300. 'I should have taken the fucking bus', I think to myself, 'I should know this by now!'. The wind whips up the dust around me as my suitcase weaves in between the rubbish, tear gas canisters and cars that battle for space in front of Israel's prime spatial weapon of choice. I start to panic. I usually take the bus, taking full advantage of that guilty tourist privilege of being able to bypass the Israeli apartheid. I pace in front of the checkpoint and phone a friend who calms me down enough to join the queue of Palestinians at the checkpoint. Today, the checkpoint isn't so busy, yet still the metal turnstiles of the checkpoint lock the occasional person in their claws. The reaction of the trapped is either one of defeated silence or shouting. I look on in sympathy, and feel the weight of the place press on my skull. It feels like an abattoir – people penned in between metal bars and chicken wire meshed roofing – or as if the film Saw *and Luton Airport had combined. I get called up from the crowd and show my passport – my (shameful) right of return. The soldier looks at it and laughs as she takes a sip of coffee.*

2pm: I take up a stool at a busy café in central Tel Aviv. I have arranged to meet up with a newfound English-Israeli friend of mine for a coffee. A warm Mediterranean breeze wafts into the café at regular intervals, lapping around me at just the right temperature with the salty scent of the sea. My Americano arrives, again, just how I like it. Young, beautiful Israelis tap away on their Macbook Pros surrounded by notepads full of study notes while Rihanna's 'Work work work work' thumps gently in the background. I pick up my cup, looking out onto the sunny plaza filled with children playing in the square's fountain and smile incredulously. The setting is lovely but absurd. I only travelled 55 km, yet in just a few hours, I travelled further than these young Israelis will ever go.

Conclusions

Recently, at a tourism conference in Copenhagen, I presented another autoethnography that I, along with two others, have been working on which explores the subjective experiences of tourism within contested space – in this case focusing on the town of Bethlehem, Palestine (see Shepherd *et al.*, 2020). I presented a particularly emotive section of my text that dealt with my horror, disgust and self-questioning that arose from walking around Bethlehem's Separation Wall and refugee camp as part of a guided tour. To aid my listeners, I provided pictures of the wall and contrasted my 'tourist' experience of Bethlehem with the Bethlehem of my childhood by playing the Christmas carol 'O Little Town of Bethlehem' in the background. The idea was to provide a powerful sense of dystopia that summarised my experience of Bethlehem and to provoke an emotional response which might spur reflection. I was nervous about it – it was pretty 'out there'. As a young academic with little to show, I was anxious to see how the establishment would take it – would they label me 'self-indulgent'? Would they question my spilling of the beans? Would they ask the classic question, 'How is this even science?'[18] Yet when I finished, I was pleasantly surprised to hear mostly comments of encouragement and intrigue, particularly, may I add, from the male scholars. The men in the room started talking about how they would have loved to try autoethnography when they were doing their doctoral theses and how they had been discouraged from subjectivity at the time due to the supremacy of objectivist and positivistic ways of working. Throughout the conference, nearly all the men I spoke to seemed intrigued, and dare I say keen, to taste the forbidden fruits of subjectivity for themselves, to unleash the Pandora's box of the epistemology of emotion (Denzin, 1997) that they had self-consciously kept shut.

Perhaps this tells us something reassuring about masculinity – that its hold is cracking. It reminds us that being male is not some 'innately fallen state' (Webb, 2017: 88), but that men have grown up saddled with the expectation not to talk about emotions, and so they need encouragement to brave engaging with them. As an academy, we should therefore engage with methods that enable the hidden or subjugated realms of knowledge (Foucault, 1994) of tourism to emerge and, in the process, hopefully encourage men to see that it is okay to cry.

> And as I sat there, hollowed out by the spectacle of the occupation, yet filled anew by the kindness of the Palestinian staff, I watched a repeated clip from the documentary film *Five Broken Cameras*. A Palestinian is shot next to an

olive tree. His friend howls at the soldier 'What was that for?!' The soldier seemed not able to answer. And I couldn't stop the tears welling up in the bottoms of my eyes. (From my personal diary, February 2018, at the Walled Off Hotel, Bethlehem)

Notes

(1) This chapter was in many ways inspired by narrative accounts that deal with the consequences of contemporary masculinity. I consciously decided to imbue the text with references to both Webb's *How Not To Be A Boy* (2017) and Grayburn's *Boys Don't Cry* (2017) in order to purposefully blend 'sanctioned' and 'unsanctioned' knowledge of masculinity.
(2) A banal, meaningless phrase that explains away aggressive behaviour as just part of boyish nature. It 'teaches children such actions are endemic to masculinity' (Meyer, 2014).
(3) In *How Not To Be A Boy* (Webb, 2017: 47), Webb aptly tell us that early masculinity stresses two crucial understandings: 'A) the Sovereign Importance of Early Homophobia, and B) the Paramount Objective of Despising Girls'.
(4) Condry and Condry demonstrated back in 1976 that girls and boys were subject to gender socialisation from infancy. Their study compared the reactions of boys and girls to a jack-in-the-box. A group observed the infants and were asked to comment on the infants' reactions. When the group was told the infant was a boy, they described the infant's reaction as 'anger'. When they were told it was a girl, they said the infant was scared.
(5) Despite the problematics of using the terms 'West' or 'Western', I use them for convenience sake to encompass the 'Anglo-Saxon' cultures of North America, Northern Europe, Australia and New Zealand – post-industrial societies where conventional masculinity has been threatened by the 'feminisation of the work place' (Raisin, 2017), and thus a perceived need to reaffirm what being male actually means.
(6) It is almost taken for granted nowadays that boys and men should not really wear pink. Historically, however, this is a new trend. Paoletti (2012) demonstrated in her book, *Pink and Blue: Telling Boys from the Girls in America*, that before WWII boys were expected to wear pink and girls blue. In the 19th century, both sexes actually wore white.
(7) Auster and Mansbach (2012) demonstrated how the Disney Store continues to divide toys based on colours and characteristics, with action toys and vehicles, for example, being labelled as for 'boys only' whereas cosmetics and jewellery were for 'girls only'.
(8) Prince Charming is not even given a name, for Pete's sake!
(9) From now on I shall often refer to the masculinity described above as 'The Trick'. I feel it is a fitting way to label the 'toxic' masculinity that has entrapped so many men over recent generations. It also means I am careful not to generalise *all* masculinity as negative and damaging – indeed, masculinities are plural (Aboim, 2010) and can be performed in different ways by different people in different societies. The Trick therefore speaks about a toxic 'Western' masculinity which has championed a detached, strong, commanding male.
(10) Harari (2014) argues that it is humankind's ability to cooperate in large numbers through the sharing of and belief in common stories – whether that be, for example, religion, money or nationalism – that makes humans the most powerful species on earth.
(11) One of the most valuable moments of my short academic career was when my supervisor and boss said to me that even though we like to think we can leave our emotional baggage at the door when we come to work, we bring that suitcase with us into our offices and into the field. More importantly, he stressed that this is ok.

(12) In fact, for some within the wider social science family, the pill was too hard to swallow. Elaine Campbell revealed how autoethnographers routinely received online abuse from male scholars who saw autoethnography as the preserve of 'self-obsessed c**ts' or 'How idiots get PhDs' (Campbell, 2017: 3).
(13) The Walled Off Hotel, which offers 'the worst view in the world', seeks to deliberately 'comfort the disturbed and disturb the comfortable' (The Walled Off Hotel, 2017). It is quite clear that British visitors are the target audience of many of the artworks in the hotel that deal with Britain's colonial legacy in the region, particularly Britain's signing of the Balfour Declaration and its disastrous consequences for Palestinians. I explore this experience autoethnographically in Shepherd *et al.* (2020).
(14) Autoethnography is usually an *ex post facto* decision. It is an attempt to make sense of a lived experience (Brown, 2015; Ellis *et al.*, 2011; Noy, 2007). It is, after all, a reflexive engagement – reflecting back over the past.
(15) This refers both to the Bantustanisation of the West Bank following the Oslo Peace Agreement of 1993 and the resultant control areas A, B and C, but also to the architecture of the occupation, of which spatial complexity is a key component (See Pullan, 2013; Weizman, 2017).
(16) This can take the form of what are called Dual Narrative Tours. During these tours, tourists will be guided round a divided city such as Jerusalem or Hebron and gain a sense of the differing narratives both sides have about that place. Schneider (2019) explored the role these tours have to play in creating increased understanding about the Israeli-Palestinian context among tourists.
(17) It is for this reason, for example, that tourists are encouraged by solidarity groups to help with the Palestinian olive harvest, as Israeli soldiers are less likely to disrupt the harvest if tourists are present.
(18) To which I would always quote Foucault (1994: 23): 'What type of knowledge do you want to disqualify in the very instant of your demand: "Is it a science?" Which speaking, discoursing subjects – which subjects of experience and knowledge – do you want to "diminish" when you say: "I who conduct this discourse am conducting a scientific discourse"?'.

References

Aboim, S. (2010) *Plural Masculinities: The Remaking of the Self in Private Life*. Abingdon: Routledge.

Anderson, L. (2006) Analytic autoethnography. *Journal of Contemporary Ethnography* 35 (4), 373–395.

Auster, C. and Mansbach, C. (2012) The gender marketing of toys: An analysis of color and type of toy on the Disney Store website. *Sex Roles* 67 (7–8), 375–388.

Balvin, N. (2017) What is gender socialization and why does it matter? *UNICEF.org*, 18 August. See https://blogs.unicef.org/evidence-for-action/what-is-gender-socialization-and-why-does-it-matter/.

Beeton, S. (2016) The self as the data: Autoethnographic approaches. Paper presented at TTRA International Conference, Vail, CO, June.

Best, G. (2017) Media memories on the move: Exploring an autoethnographic heritage of automobility and travel. *Journal of Heritage Tourism* 12 (1), 52–66.

Botterill, D. (2003) An autoethnographic narrative on tourism research epistemologies. *Loisir et Société/Society and Leisure* 26 (1), 97–110.

Brown, L. (2015) Treading in the footsteps of literary heroes: An autoethnography. *European Journal of Tourism and Hospitality Research* 7 (2), 135–145.

Buda, D. (2015) *Affective Tourism: Dark Routes in Conflict*. London: Routledge.

Buda, D. and McIntosh, A. (2012) Hospitality, peace and conflict: 'Doing fieldwork' in Palestine. *Journal of Tourism and Peace Research* 2 (2), 50–61.

Burlingame, K. (2019) Where are the storytellers? A quest to (re)enchant geography through writing as method. *Journal of Geography in Higher Education* 43 (1), 567–570.

Campbell, E. (2017) 'Apparently being a self-obsessed c**t is now academically lauded': Experiencing Twitter trolling of autoethnographers. *Qualitative Social Research* 18 (3), 1–19.

Chaudhry, L.N. (2009) Forays into the mist: Violences, voices, vignettes. In A.Y. Jackson and L.A. Mazzei (eds) *Voice in Qualitative Inquiry: Challenging Conventional, Interpretive, and Critical Conceptions in Qualitative Research* (pp. 137–163). London: Routledge.

Condry, J. and Condry, S. (1976) Sex differences: A study of the eye of the beholder. *Child Development* 47 (3), 812–819.

Davenport, L. and Hall, J. (2011) To cry or not to cry: Analyzing the dimensions of professional vulnerability. *Journal of Holistic Nursing* 29 (3), 180–188.

Delamont, S. (2007) Arguments against auto-ethnography. Paper presented at the British Educational Research Association Annual Conference, London, September.

Denzin, N. (1984) *On Understanding Emotion*. San Francisco, CA: Jossey-Bass.

Denzin, N. (1997) *Interpretive Ethnography: Ethnographic Practices for the 21st Century*. Thousand Oaks, CA: Sage.

Denzin, N. and Lincoln, S. (2000) *Handbook of Qualitative Research* (2nd edn). Thousand Oaks, CA: Sage.

Doty, R.L. (2010) Autoethnography – making human connections. *Review of International Studies* 36, 1047–1050.

Ek, R. and Larson, M. (2017) Imagining the alpha male of the tourism tribe. *Anatolia* 28 (4), 540–552.

Ellis, C. and Bochner, A.P. (2006) Analyzing analytic autoethnography: An autopsy. *Journal of Contemporary Ethnography* 35 (4), 429–449.

Ellis, C., Adams, T. and Bochner, A.P. (2011) Autoethnography: An overview. *Forum: Qualitative Social Research* 12 (1), Art. 10.

Emerald, E. and Carpenter, L. (2015) Vulnerability and emotions in research: Risks, dilemmas, and doubts. *Qualitative Inquiry* 21 (8), 741–750.

Figueroa-Domecq, C., Pritchard, A., Segovia-Pérez, M., Morgan, N. and Villancé-Molinero, T. (2015) Tourism gender research: A critical accounting. *Annals of Tourism Research* 52, 87–103.

Foucault, M. (1994) Two lectures. In M. Kelly (ed.) *Critique and Power: Recasting the Foucault/Habermas Debate* (pp. 17–47). Cambridge, MA: MIT University Press.

Gilbert, K. (2000) *The Emotional Nature of Qualitative Research*. Boca Raton, FL: CRC Press.

Grant, A. (2010) Writing the reflexive self: An autoethnography of alcoholism and the impact of psychotherapy culture. *Journal of Psychiatric and Mental Health Nursing* 17 (7), 577–582.

Grayburn, T. (2017) *Boys Don't Cry*. London: Hodder & Stoughton.

Harari, Y. (2014) *Sapiens: A Brief History of Humankind*. London: Harvill-Secker.

Himmelstein, M. and Sanchez, D. (2016) Masculinity in the doctor's office: Masculinity, gendered doctor preference and doctor–patient communication. *Preventative Medicine* 84, 34–40.

Kaufmann, V., Bergman, M.M. and Joye, D. (2004) Motility: Mobility as capital. *International Journal of Urban and Regional Research* 28 (4), 745–756.

Kavoori, A. (2018) The most peaceful place on Earth. *Cultural Studies – Critical Methodologies* 18 (2), 116–118.

Meyer, E. (2014) The danger of 'boys will be boys'. *Psychology Today*, 14 September. See https://www.psychologytoday.com/gb/blog/gender-and-schooling/201403/the-danger-boys-will-be-boys.

Miraftab, F. (2004) Can you belly dance? Methodological questions in the era of transnational feminist research. *Gender, Place & Culture* 11 (4), 595–604.
Munar, A.M. (2018) Researching in a men's paradise: The emotional negotiations of drunken tourism fieldwork. In B.A. Porter and H.A. Schänzel (eds) *Femininities in the Field: Tourism and Transdisciplinary Research* (pp. 170–184). Bristol: Channel View Publications.
Noy, C. (2007) The poetics of tourist experience: An autoethnography of a family trip to Eilat. *Journal of Tourism and Cultural Change* 5 (3), 141–157.
Paoletti, J.B. (2012) *Pink and Blue: Telling the Boys from the Girls in America.* Bloomington, IN: Indiana University Press.
Patton, M. (2004) Autoethnography. In M.S. Lewis-Beck, A. Bryman and T. Futing Lao (eds) *The SAGE Encyclopedia of Social Science Research Methods* (pp. 46–48). Thousand Oaks, CA: Sage.
Porter, B.A. and Schänzel, H.A. (eds) (2018) *Femininities in the Field: Tourism and Transdisciplinary Research.* Bristol: Channel View Publications.
Pritchard, A. and Morgan, N. (2000) Privileging the male gaze. *Annals of Tourism Research* 27 (4), 884–905.
Pritchard, A. and Morgan, N. (2007) De-centering tourism's intellectual universe, or traversing the dialogue between change and tradition. In I. Ateljevic, A. Pritchard and N. Morgan (eds) *The Critical Turn in Tourism Studies: Innovative Research Methodologies* (pp. 11–28). London: Routledge.
Pritchard, A., Morgan, N. and Ateljevic, I. (2011) Hopeful tourism: A new transformative perspective. *Annals of Tourism Research* 38 (3), 941–963.
Pullan, W. (2013) Conflict's tools: Borders, boundaries and mobility in Jerusalem's spatial structures. *Mobilities* 8 (1), 125–147.
Raisin, R. (2017) Men or mice: Is masculinity in crisis? *The Guardian*, 6 October. See https://www.theguardian.com/inequality/2017/oct/06/men-or-mice-is-masculinity-in-crisis-ross-raisin.
Reinharz, S. (1997) Who am I? The need for a variety of selves in the field. In R. Hertz (ed.) *Reflexivity and Voice* (pp. 3–21). Thousand Oaks, CA: Sage.
Samaritans (2019) *Suicide: Facts and Figures.* See https://www.samaritans.org/about-samaritans/research-policy/suicide-facts-and-figures/.
Schneider, E. (2019) Touring for peace: The role of dual-narrative tours in creating transnational activists. *International Journal of Tourism Cities* 5 (2), 200–218.
Sheller, M. (2014) The new mobilities paradigm for a live sociology. *Current Sociology* 62 (6), 789–811.
Shepherd, J., Laven, D. and Shamma, L. (2020) Autoethnographic journeys through contested spaces. *Annals of Tourism Research* 84. Advanced Online Publication.
Shondell Miller, D.M. (2008) Disaster tourism and disaster landscape attractions after Hurricane Katrina: An autoethnographic journey. *International Journal of Culture, Tourism and Hospitality Research* 2 (2), 115–131.
Sparkes A.C. (2002) Autoethnography: Self-indulgence or something more? In A.P. Bochner and C. Ellis (eds) *Ethnographically Speaking: Auto-ethnography, Literature and Aesthetics* (pp. 209–232). Walnut Creek, CA: Alta Mira Press.
The Walled Off Hotel (2017) Instagram page. See https://www.instagram.com/walledoffhotel/.
Thien, D. (2005) After or beyond feeling? A consideration of affect and emotion in geography. *Area* 37 (4), 450–456.
Thurnell-Read, T. (2011) 'Common sense' research: Senses, emotions and embodiment in researching stag tourism in Eastern Europe. *Methodological Innovations* 6 (3), 39–49.
Wall, S. (2006) An autoethnography on learning about autoethnography. *International Journal of Qualitative Methods* 5 (2), 146–160.

Wang, Y., Hunt, K., Nazareth, I., Freemantle, N. and Petersen, I. (2013) Do men consult less than women? An analysis of routinely collected UK general practice data. *BMJ Open* 3 (8), 1–8.

Waterton, E., Staiff, R., Bushell, R. and Burns, E. (2018) Monster mines, dugouts, and abandoned villages: A composite narrative of Burra's heritage. *Journal of Heritage Tourism* 14 (2), 85–100.

Webb, R. (2017) *How Not to Be a Boy*. Edinburgh: Canongate Books.

Weizman, E. (2017) *Hollow Land: Israel's Architecture of Occupation* (new edn, originally published 2007). Brooklyn, NY: Verso.

Wilson, E. and Hollinshead, K. (2015) Qualitative tourism research: Opportunities in the emergent soft sciences. *Annals of Tourism Research* 54, 30–47.

Wright, R.K. (2018) Doing it for Dot: Exploring active ageing sport tourism experiences through the medium of creative analytical practice. *Journal of Sport and Tourism* 22 (2), 93–108.

2 When Is a Hegemonic Male Not a Hegemonic Male? Personal Reflections of a Tourism(ish) Researcher

Neil Carr

The hegemonic male, the symbol of unbridled power, is the demi-god of society. There is nowhere he cannot go and nothing he cannot do unless he does not wish to do it because it is irrelevant to him or beneath him. That there are males in society who believe this to be true of and for themselves is, alas, beyond dispute. They are the unthinking meatheads of the world. This chapter is not meant to analyse this delusion, but to provide a reflective personal discussion of an apparently hegemonic male's explorations of gender, feminism, children and families, welfare, well-being and sex within the context of tourism. The chapter reflects on the influences of such a journey on personal identity as 'male' and 'hegemonic male' and how the physical body can define and constrain how others perceive us. The chapter explores the barriers that have to be faced and overcome by males working in these areas. Ironically, in overcoming such barriers the chapter questions whether I am actually the very thing I have railed against, the hegemonic male against whom no doors can be closed. The chapter ends by suggesting, perhaps in hope, but hopefully not a forlorn hope, that this cannot be the case as while breaking down doors this male has constantly been racked with self-doubt and fear (including in the 'baring of all' that this chapter represents) – two characteristics most assuredly not associated with the hegemonic male.

Introduction: A Lifetime of Being Male (and Hegemonic?)

At long last, we live in an era where there is at least some acceptance of the idea that trying to divide people into binary, anatomically grounded sex categories is at best a flawed exercise. Yet old social orders and habits die hard and, as a result, it is still very common to see us being divided according to whether we are male or female. It is into the former category that I have been positioned and, through desire or social conditioning, am content to ascribe to now. Further to that, I carry many of the trappings of that most targeted of species, the hegemonic male. This male is a white (tick), Anglo-Saxon (tick), protestant (tick, thanks to parents who through social conditioning felt it necessary to give a baby to a priest to be sprinkled with some cold water before he could say no). This creature is not only male but also overtly and covertly heterosexual (tick – doubly underlined because I have evidence to prove it, i.e. wife and three children – the latter not only proclaiming my heterosexuality[1] but also my virility). Furthermore, this creature encapsulates and presents to the world all that society has defined as the dominant features of masculinity. As such, in its physical appearance it is muscular (tick, in a slightly scrawny way). In the portrayal of itself to the world it is confident (tick, so I am told), and in its own head it is the lord of all that it deliberate (we will come back to this later). This is not just the hegemonic male; it is the alpha male. It knows nobody will challenge it and fully accepts that all should kneel in awe of its awesomeness (again, we will come back to this).

If we dig a little deeper into the community of the hegemonic male, it is possible to begin to discern that they are not all one and the same. There are the wannabe hegemonic males. You can find these every weekend on the streets of many cities, with their chests puffed out and their biceps rippling. They exude an almost palpable taste and smell of masculinity, itching for a fight or a sexual conquest and the resultant opportunity for them to assert their dominance over the 'other' and to be seen to do so. There are also the more covert hegemonic males. These hide behind a veneer of social acceptability but are just as prone to a belief that they can take whatever they want and do whatever they want, safe in the notion that their own greatness will absolve them of any blame that would be attached to the same actions carried out by mere mortals. Both of these types of hegemonic male are as identifiable in the corridors of academia as they are in the wider society. Readers are invited to reflect on their own experiences to identify these people.

These hegemonic males come across rather badly. They are arrogant, conceited and, more than that although more crudely put, they are meatheads whose decisions and subsequent behaviour appear to be driven more from an area to the south of their waistlines than from their brains (why else do we call them dickheads?). It really is no wonder that set against this portrait so many groups have, in the past 20 or more years, sought to castigate these creatures and post them into the fires of hell.

So am I one of these creatures? Well, as already noted, in many ways the answer must be yes. Other evidence to level against me could include my professorship and my position as a head of department, those positions of authority that are so heavily biased in favour not just of men but of hegemonic males. There is also my ultra-competitive streak that sees me out on the harbour in my home city in my single scull (rowing, for the uninitiated) and racing, or at least attempting to race in my own mind, anything else on the water. Just like myself, the hegemonic male hates to lose.

When I began my journey as a researcher (we will note the beginning as when I began my Honours dissertation in 1993 though that is arguably an arbitrary date), I did not have a few of the trappings of the hegemonic male that I have now gathered along the way. I did not have any children, although I had recently met the woman who would go on to become my wife and mother of our children. I did not have a professorship and nor was I head of department at a university, although I had already been president of the University of St Andrews rowing club and Athletic Union. However, I did have plenty of things that identified me as a hegemonic male, whether I liked it or not. This included the blissful ignorance of the hegemonic male that negates the need to ever question who or what it is or where it fits. Instead, of course, the hegemonic male is not a creature of doubt; rather it simply is, and knows that, if it cares to think about it at all, which it probably does not.

While it is undoubtedly fun to knock the hegemonic male – indeed it is today a socially sanctioned sport – this is not the aim of the chapter. Rather, it is meant to provide a self-reflection on the hegemonic male and his gaze in tourism research. In the process, it questions my 'me-ness'. It explores the barriers to researching in certain fields faced by the hegemonic male. The chapter also provides an opportunity to explore what, if any, influence my research has had on my positionality in regard to the hegemonic male.

Researching beyond the Hegemonic Male Gaze

For my first foray into research, my Honours dissertation was a tame affair. True, it looked at tourism research in an ultra-traditional geography department (University of St Andrews). However, I really did what any sane student would do and played it safe. The rationale behind this decision was that the dissertation is about getting the grades to get the degree. Then I began, inadvertently and certainly not by personal design, to grow a little. A PhD proposal was beaten out of my head and at the same time nursed into existence and taken lovingly and in fear on its first tentative steps into the world. The former suggests a strong hegemonic male while the latter suggests something contradictory. The reality is that the truth was not somewhere in between these two; rather, it *was* these two. We will return to this apparent conflict as we wander through this chapter. However, it is worth beginning to ponder if this contradiction was a sign that something lay beyond the hegemonic male.

Life took one of those important crossroads that are only recognised as such much later when one of my supervisors-in-waiting urged me not just to look at youth behaviour in the holiday experience, but to explore the gendered nature of such behaviour as well. The, in retrospect, hegemonic male thinking was that gender was at the time one of the 'in' topics and that putting it in the proposal enhanced my chances of gaining a scholarship and as a result being able to afford to do the PhD. Was this my lightbulb moment on the road to Damascus? No, but on reflection it certainly set me on that road, or maybe I had always been on it, but had never known it or cared to look. Whichever the case, I delved with the enthusiasm of a newbie PhD student, complete with scholarship, into the literature to find the rocks on which to begin constructing my thesis.

Here was that apparently hegemonic male falling in love with the works of the leading gender and feminist researchers and writers in leisure studies at the time. I had quickly abandoned looking in the tourism literature and instead had stumbled on the leisure field, with its rich history of critical thinking that while often recognising the reality of the leisure industry is rarely beholden to it. Sue Shaw, Rosemary Deem, Karla Henderson, Betsy Wearing and many others in leisure studies lit a flame in me. Running back to my geographical roots, Linda McDowell, Doreen Massey and Gill Valentine, among others, fanned the flame still further.

Was I in full reflexive mode by then, in the late 1990s? No, but I was busy delving into readings about feminism for the first time and this would later be augmented by visits to gender and leisure conferences and meetings about gender at more wide-ranging conferences. I cannot remember with any clarity if I was the only male at such conferences or meetings but I can say that I was most assuredly in a minority. Nevertheless, oh what an opportunity, to read about the subjugation of people and the route to empowerment through self. Feminism, bring it on please – I thought, and still think. Only after the conferences and meetings did I reflect on the fact that the women at these events had not just suffered me but, I like and hope to think, accepted me – a (hegemonic) male in their midst. I am so lucky that they did, for I met not just those I had idolised from their writings but new people like Deb Bialeschki, Heather Gibson and Cara Aitchison, among many others. In my wanderings in this area, I talked of empowerment, of women entering leisure spaces in defiance of social teachings on gender and personal fear. I also talked of the fact that while men may have entered leisure spaces it did not mean that they did so without fear. In other words, although I may never have put it so eloquently, these young men and women were engaged in their own feminist struggle, seeking the right to be free. Is this where my obsession with the link between leisure and freedom came from or is it a more deep-seated desire for personal freedom? Another question to add to the fire of self-doubt.

Reflecting back, did I ever stop to think whether attending conferences and other meetings focused on gender was an appropriate behaviour? No, I did not, and that in itself could be said to be a product of my position as a hegemonic male. Yet I would argue that I did not enter those spaces because I felt I had a God-given right to do so but rather because it seemed the natural place for someone studying gender and leisure to go. It really was only when I went to these events that I realised how few men were studying aspects of gender. This struck me, and still strikes me, as very odd indeed. Women may well be perfectly capable of researching gender without men, but surely men should be as interested in this research field as women obviously are. Yet like the young people struggling to access leisure spaces, when I went to gender-focused conferences and meetings it did not mean that I did so without fear or trepidation. Even today, I attend each conference and give each presentation with deep-seated unease, a feeling that I do not belong, that I am not worthy. This self-doubt will be returned to later in the chapter, but for now it is sufficient to recognise that it has no place in the headspace of the hegemonic male.

Yet gendered leisure (or tourism – in another sign of self-doubt, I never have figured out whether I am a leisure or tourism researcher) could not hold my attention for long. Ironically, I became bored and frustrated with the male–female binary that was still dominant at the time. Masculine and queer studies were just emerging, at least on my radar, but rather than delving deeply into those I wandered into child and family studies. This shift was driven by the birth of my first child and the whole rewiring of life that this created. Unknowingly, I think, I brought feminism along for the ride and began to criticise the existing work for not seeing children for what they are, active social agents. I railed against the disempowered position of children in tourism and leisure studies, seeking to give agency to their voices. Once again, I had landed in a research field dominated by women. The health, or otherwise, of this situation has been questioned by Schänzel and Carr (2016).

The next stop on my research journey was dogs. Why? Because I became a dog owner, and later friend, for the first time in 2001, and by 2002 I found myself railing against perceived social injustices regarding the access of dog owners and their pets to leisure experiences. Once again, feminism had come along on the journey with me. From dogs, I expanded to encompass the entire animal kingdom and ran through animal rights and welfare to the concept of human obligations to animals (big thanks are owed to Don Broom for introducing me to that concept). While the focus may have begun with humans facing barriers to human leisure and tourism experiences with their animals, it changed to thinking about an even more discriminated-against group, the animals.

While youth, children and families, and animals may have risen (and probably in the case of the first two fallen) as my research focus as time has moved on, sex has periodically popped its head up as a focus of my intellectual attention (is an intellectual hegemonic male a contradiction?). Why did I look into this dark corner of tourism and leisure research? Was it, as some researchers in the field of sex have worried their colleagues might think, because of a desire for sexual stimulation? This suggestion, arguably borne of a mixture of xenophobia and ignorance, has been strongly argued against elsewhere (Bullough & Bullough, 1977; Reichert, 2003). I would simply say no. Instead, the driver has been what it always has been, no matter the focus, to shine a light into dark corners and in the process maybe, just maybe, to help to improve things, even just a little bit. Yet I hear and identify with the discomfort identified by Munar (2018) when she feels compelled to write about her (sexualised) body in relation to her research on hypersexualised tourism spaces.

It is clear that my research has followed not just my interests but also my lifecycle. The two are, and always have been, intimately bound together. This is undoubtedly a highly privileged position to be in, while at the same time blurring the divide between work and non-work. It may be best here to turn to my 17-year-old daughter to try and see this through a gendered lens. She has been in full-time ballet training for several years and is devoted to classical ballet. On the cusp of attempting to pursue a professional ballet dancing career, she has all the grace, poise, elegance and beauty normally associated with ballet, even to my uncultured and biased eye. Yet behind this apparent femininity there is a school kid who loves the sciences – maths and chemistry in particular. There is also a core of steel to her, necessary to the physical and emotional toll that is charged against anyone reaching for the heights of their chosen profession. This academic focus and the steel within her could be, and have traditionally been, labelled as masculine. Yet femininity and masculinity are merely labels; they are not who we are. Just as with my daughter, my research interests in gender, children and families could be said to be feminine while my addiction to rowing could equally be labelled masculine. My research and everything else in my life is simply part of what I am and femininity or masculinity is, in oh so many ways, irrelevant. Before anyone tries to do so, that is also not an excuse to try to pin androgeneity or any other label on me.

Maybe it was that I had at long last grown up enough, or maybe I had accumulated enough 'stuff', or maybe I managed to find that most rare of academic commodities in the 21st century, free time. Whichever the case, it was time to sit down in the early teens of the new century and think back about feminism and my raging against the machine and join the dots between my seemingly disparate research streams. This brought me back to the question of freedom and its related concept of power. They exist on the route of feminism but I drove through that, not entirely satisfied, and went on to look at the various dead French philosophers, rejecting them all in turn as interesting, but not interesting enough. The dead Germanic philosophers followed and seemed to be saying things about freedom in a much better way. The dead Russian philosophers were also worth a look but again came up, in the mind of this individual at least, a little short. I found some great thoughts in less traditional places, including the teachings of such wonderful people and animals as the Dalai Lama, Gerald Durrell, Desmond Tutu, Terry Pratchett, Dog[2] and Calvin and Hobbs.

Reflecting on this list, I find it disconcerting that it is entirely filled with males – even the cartoon characters and the authors who created them. Is this a reflection of the dominant presence of men in the public sphere? Is it a subconscious leaning of my own as a male? Yet they are such a disparate group that to say they can simply be grouped together based on anatomy does them and what they stand/stood for a gross disservice. I think this leaves them simply as a group of individuals who have been influential to my thinking and who just happen to be males. To do anything more seems like self-indulgent psychoanalysis.

Whatever the shortcomings of all these authors or, possibly more accurately, the shortcomings of me as reader and my consequent inability to understand what they were trying to tell me, they all had a very important and powerful message to tell. This was that subjugation of the individual, human or not, is not acceptable and that we have obligations to help the other, but also that to truly escape subjugation is a journey that only the subjugated can take; we cannot give freedom, although we can help in its gaining. Rather, freedom must be taken. This is heady wine indeed. It links back to feminism, but is more than just feminism. What has all this taught me? That power in and of itself is useless and worthless. That freedom, even for a social animal, is priceless but that it can only ever be taken and is worthless in and of itself. Rather, with both freedom and power come obligations to others. Because of the latter, we must use the former to help others, whoever we or they are. It is also armed with this understanding of freedom that, as mentioned earlier in this section, I find the power to cast off social constructions of femininity and masculinity, and any other label, and just be me.

Barriers to the Hegemonic Male Researcher

A hegemonic male, the alpha creature of society in general and academia in particular, is surely meant to be able to go anywhere he desires and do whatever research he wishes. Yet barriers clearly exist that mean that this is not necessarily the case. I found this early on as a PhD student. It was easy enough to talk to young people, irrespective of whether they were male or female, about their holiday experiences. However, understanding what they were saying from their perspective was, on first examination, another thing entirely. I had arrived at long-debated questions in the social sciences that have stretched themselves beyond the narrow confines of positivism of how to understand the 'other'. Many academics have faced this issue but I am indebted to the work of Sue Smith (1993) on the Beltane festival in Scotland, who talked of insiders and outsiders in a society and the issue of positionality. This barrier of the 'other' is only a barrier in our mind and the inward looking, conservative mindset of society. Rather, while I may never be able to see the world as the other does, I can at least always reflect upon how the world is seen through the eyes of the other. Such an approach does not so much demand a level of empathy, although that may well help, as a willingness, an open-mindedness, to see things from the perspective of the other. In this way, the other becomes viewable as everyone and anyone. Just because I may have more personal traits in common with some people than with others does not mean that they are the same as me. Consequently, I have learned that no matter who I am looking at, talking with, researching with, they must always be seen as the 'other'. In this way, there is fundamentally no difference between conducting research with, as opposed to experimental research on, humans or non-human animals (see Carr, 2014, for more discussion of the issues associated with conducting research with the non-human other).

In the age in which we live the single male has often been cast as deviant if he enters certain spaces from which he is 'normally' absent or excluded by default if not in law. Such spaces can include theme parks, zoos, cinemas when they are showing films targeted at children, and school playgrounds. A single male entering such spaces must at a minimum deal with the gazes of those 'allowed' in such spaces and those who police such spaces. These include parents, teachers and children. The implied view here is that a single male in these spaces is a potential paedophile. Personally, it has only ever felt 'safe' or comfortable to be in a theme park or zoo with my children and/or wife. By myself I feel naked, exposed and less than welcome. In this way, my wife and children have unwittingly enacted the bodyguard role Munar's (2018) husband was given by her when she invited him to accompany her on her visits to the hypersexualised Bierstrasse area of Palma de Majorca, Spain. The gaze of society is brought fully into focus today by the very serious demands placed on researchers, not just to be ethical in all that they do, but to undergo police checks if they are working with children. I fully understand the depressing need for such checks but feel they are, nonetheless, intimidating (should a hegemonic male ever feel intimidated?) and this in itself can be a barrier to research.

What to do then when you are attempting to conduct research with children about their and their families' holiday experiences? My own solution was to hire a female research assistant. She, in the sexist world we still live in, had no problem entering primary schools to talk to children about their holiday experiences. In the case of my own children and family, I was entirely comfortable researching them (see Carr, 2011, for a discussion of all the issues associated, rightly or wrongly, with conducting social science research with your own children and family). Yet both methods speak of privilege: in the former case it is the privilege of having the money for a research assistant; in the latter case, it is the privilege of having children of my own. What of male researchers who have neither but wish to conduct research with children and their families?

The first time I attempted to explore sex in the research arena was as a tiny part of my PhD thesis. To me it seemed logical to ask young people about sex as part of a wider study on their holiday behaviour. It was by no means a focus of the research, but the reaction of my supervisors to the sight of that little word in my questionnaire the first time they saw it sticks in my memory. I cannot recall the details, but as I dig out the old files to look at the initial versions of my survey I can see that sexual activities is in there, but by the time we reach the final version it has been airbrushed out in the face of concern about the consequences of asking people about their sexual experiences. I had not read the material in the academic press, noted earlier in this chapter, that seeks to explain why, relative to its central position in life, sex in its infinite variety is so poorly studied by academics. If I had, I would have noted the cautionary voices that urge academics, in particular, to avoid looking at sex in their research for fear of forever being labelled a pervert and damaging an embryonic academic career. Were my supervisors trying to provide similarly cautionary advice? Quite possibly. Yet just like my children and I will always keep picking at a scab, after my PhD I quickly returned to the issue of sex in relation to tourism behaviour. It was buried in a multinational survey I ran about the holiday behaviour of university students. Again I ran into a roadblock when some of those who volunteered to help distribute the survey at their universities chose to remove the sexual references from the version they distributed. Only after the event did they cite the sensitivity of the topic in their country as the reason for doing so. Really, I cannot blame them and instead can only thank them for the data they collected. Yet this survey was only a very tame toe-dip into the world of research about sex and I left it there as I wandered off into the other research areas noted earlier in this chapter. Yet I have returned to the issue several times, each hopefully with a little more maturity and open-mindedness. This needs situating within the reality that I am a child of the 1970s, raised in a conservative household in the northeast of England – not exactly fertile ground for liberal thinking and being. I was no longer a newly minted young academic but instead one with a reputation for doing the research he wished (acting like a hegemonic male!). Looking at the nature of sex shops was both fun and illuminating. It showed how such sites can be empowering and enslaving, but also that researching in this field can be liberating for academics and, infinitely more importantly, beneficial for everyone whose lives sex touches and can touch. It drove me on to realise that hiding issues in the dark simply because society does not wish to recognise reality is not a solution; indeed it only helps problems to continue and even become worse.

Espousing the social benefits of research on sex is nothing new and arguably helps to legitimise such work (Carr, 2016). Yet even so, conducting research on sex in leisure or tourism requires a thick hide that is impervious to questions about the validity of the work versus the sexual motivation of the researcher. This was most recently brought to the fore of my thinking when explaining to colleagues my research visit to Amsterdam's sex museum, which really was a lovely opportunity to examine human behaviour. Yes, the images and objects on display were interesting, to varying degrees, but people's reactions to them were far more intriguing. I am sure that all those who touch upon issues of sex in their research have had their motives questioned, but it is equally clear that such questions have not stopped an array of men and women. Perhaps being a hegemonic male gives me a slight advantage in providing an impervious coating to my hide. However, plenty of women have conducted truly ground-breaking research in this field. Among them, Liza Berdychevsky is, I would suggest, the leading researcher in the leisure studies field looking consistently at sex (i.e. Berdychevsky & Gibson, 2015; Berdychevsky & Nimrod, 2017), and of course Feona Attwood (i.e. Attwood & Smith, 2013) deserves a mention. Perhaps this is an example of the point that attempting to distinguish people across a binary male–female divide is erroneous.

It was with this mindset that I most recently came across a barrier I have yet to overcome. In the process of background reading and thinking for my book entitled *Dogs in the Leisure Experience* (2014), I happened upon the concept of bestiality. I suppose I had heard the word before over the years; after all, although rarely brought into the light of day by society, it is something that does occasionally either sneak into the light or is dragged kicking and screaming into it, but it was not something I had ever paid any attention to. However, by this time in my life I was a totally committed dog lover (plutonic) and was just about to begin to expand my emergent ideas on dog welfare to the rest of the animal kingdom. I wanted to know about this. I wanted to drag it into the light of day, examine it and not bury it but destroy it. Yet it was clear that here I was stepping into a very sensitive area for even someone who was no longer an early career academic. There is hardly any academic literature on the topic and it is all very limited. Data about bestiality were simply not available. This left me frustrated that such an issue which was afflicting the well-being of an untold number of animals was being left in the dark to continue its existence rather than being dragged into the light and therefore dealt with. I was repulsed at the notion of doing any research, however small and insignificant, in this area but equally disgusted at the idea of doing nothing.

After much soul searching I did what the hegemonic male would never do. I reached out for help from, for the sake of anonymity, a person we will simply identify as someone I trust and respect. I asked about the sanity of conducting a sweep of the internet to begin to gain, no matter how tenuous, a grasp of the scale of bestiality in the contemporary era. This, I argued to myself and this person, was a necessary precursor to any potential solutions to ensuring the welfare of animals. The considered advice was not to go there, a repeat of all the advice the younger researchers have been told in the literature regarding looking at sex in their research. Have I broken this barrier yet? No. I continue to feel physically sickened at the thought of doing that research and yet, as a hegemonic male, I continue to castigate myself as a wimp for not doing it. I also am very much aware of the sensitivity of the topic and the resultant potential for harm. However, the argument goes, if I believe in animal welfare and I see an animal welfare problem I owe it to both the animals and my own beliefs to do something about it. Perhaps I will, but perhaps I am also doing so by simply writing about the issue here (and elsewhere; see Carr, 2014, 2016). I cannot say that I am happy with this little step, but perhaps it is better than no step at all and, after all, all journeys must begin with a first step.

Am I, or Am I Not a Hegemonic Male?

Behold, all ye mere mortals, I am the alpha hegemonic male. Now kneel to my needs and demands. A hegemonic male, I have suggested in this chapter, does what he wants and to him the barriers do not apply, either because he simply thinks this is the case, or because society agrees with him, or both. Reflecting back, I have, as already noted in this chapter, at least from one perspective, largely done the research I want throughout my academic career, irrespective of the barriers. Many of those who have known me have cringed at some of the topics I have looked at and the related issues I have raised at conferences, in seminars and lecture theatres, in text and at home, yet arguably like a 'good' hegemonic male I have gone ahead and done this anyway. If it has meant treading in sensitive areas and asking sensitive questions, that has not stopped me and various ethics committees have just had to deal with it.

Yet if we scrape below the surface of this view, as all good researchers surely must when going about their work, does the image of Neil as the hegemonic male hold? True, just as with many academics, a lot of the research we do is done because we enjoy it. However, as a mature academic I more and more often find myself researching (in its broadest sense which includes producing publications, editing journals, getting research grants and supervising graduate students) not simply for my own titillation and/or gain but to help others. I do not do any of it for the awards or distinctions. They are nice but they are also deeply humbling and in the process a bit embarrassing – linking to the imposter syndrome. I certainly do not do it to see my name in the press or my image on social media. In addition, as I hope I have demonstrated in this chapter, I am not the hegemonic male that is certain of who and what he is and therefore never encounters self-doubt. Rather, such doubt is a constant companion that requires me to regularly reflect on what I do. Yet to function in an academic environment, indeed a world that has arguably been shaped and dominated by hegemonic males, such self-doubts cannot be allowed to conquer the individual. Instead, in a display of hegemonic masculinity, they must be conquered, however temporarily. Have I succeeded in conquering the self-doubts? No, but I have managed to put them aside sporadically. I've talked to enough male academics, established and emerging, to know I am not alone in grappling with imposter syndrome. Rather, it is thriving out there in academia among a whole raft of people, irrespective of gender, who really have no reason to doubt themselves. The irony of imposter syndrome is that while I fervently believe this is the case for others I continue to believe it is not true for me.

Conclusion

Does the hegemonic male exist? As a social construct, I would say it most certainly does. At the same time, I would suggest few, although possibly still far too many, males fit neatly into the tightly bound definitions of hegemonic masculinity. Is a hegemonic male a bad thing, to be slain like dragons of old? If we look only at the bad side of the beast then the answer is definitively and resoundingly *yes*. However, all social constructs have a good side, even if it only allows us to assess our 'self' against a standard we wish to distance our self from. If we wipe the bogeyman from the face of the planet we may have less to fear but our ability to judge what we should, and should not, fear may be degraded. Am I a hegemonic male? I would argue that in asking this question I am clearly not. A hegemonic male would have no need for such a question; self-doubt is not in the DNA of the hegemonic male. To have any empathy, no matter how imperfect, for others, to wish to see the lives of others improved and for those humans and non-human animals to be empowered through the obligations of those in power and through themselves are not the thoughts of the hegemonic male. Indeed, such thoughts are clearly at the heart of the origin of feminism, a concept that offers so much to so many beyond the narrow confines of a binary anatomical sex divide. Yet, I clearly look like a hegemonic male, and in many ways act and even think like one. However, we are clearly today in a society where nobody should be discriminated against simply because of the way they look.

So as one who looks like a hegemonic male to the casual gaze, in the kinds of positions of power (journal editor, head of department, professor) that are supposed, in the minds of hegemonic males, to have been made for them, I offer one final thought that suggests I must not be a hegemonic male. That is, that instead of seeing power over things and people I see obligations: obligations to utilise what little power I have, because power really is mainly an illusion, to help others. This brings us neatly full circle to realise that it is not just my research that is about helping others (human and non-human); it is my entire life as an academic. However, it goes further than that; it reaches out to my entire life. It is about helping my children to be what they want to be (16-year-old daughter = a professional classical ballet dancer; 19-year-old son = a World and Olympic swimmer; eight-year-old son = an eight-year-old kid), and helping my wife in everything from the menial tasks of life to being whatever it is she wishes to be. It is also about helping all the animals I can as well as the various people whose paths I cross as I wander through life. I am not perfect. I am not Cooch (see *Footrot Flats* – it really is an educational read) or one of my PhD students, Ismail, who do everything they can to protect the lives of every animal that crosses their paths, however remotely. Neither am I Mother Teresa or Desmond Tutu. However, I do strive to do better every day. So the child that supposedly happily pulled the wings and legs off poor daddy-long-legs (I'm sorry – another trait alien to the hegemonic male) has been replaced by the adult who miserably cried, and is now weeping as he types this, as he held Snuffie, his first dog, in his arms as she died. She had been administered a lethal injection by a family friend and vet to save her from what had become the never-ending agony of cancerous growths in her spinal column. Yet there again is the hegemonic male raising its head, for it was I that in the final analysis made the decision to end Snuffie's life. Others, our vet and my wife primarily, gave advice and guidance, but the decision was mine. Best friend or hegemonic male wielding the ultimate power of life and death – you decide; personally I have come to terms with that in my own mind. I am a friend, and with that comes huge responsibilities, and one of those is making the decision of when it is best for all the animals I have and still do share my life with to end their suffering.

My final message is to all those hegemonic males out there: get your brains out of your crotch, please, although I seriously doubt the fully committed hegemonic male will ever listen. To those, like me, who may look like a hegemonic male, or have been more of one in the past, change is possible, and we do not need to be what we look like just because society says so. We are capable of empowering ourselves to change. It is empowerment, though, that is of central importance to this whole discussion. It is arguably beholden on hegemonic males to grasp the important questions that others fear to ask, to use their supposed power to help others either directly or by opening up research agendas that had previously been barricaded. Have I done enough of this? No, I think I have spent too much time captured by an academic system that demands obsequiousness to the god that is publishing and the gaining of external research income. So my next challenge, for there always is one, is to become more of an activist who uses his power, however limited it may be, to improve the well-being of others, and in so doing to empower them to empower themselves and where even this is impossible to ensure that they are guarded from exploitation. As part of the review process of this chapter I was asked by Heike Schänzel whether a mature hegemonic male is more in touch with his feminine side. I guess that in this there was the implicit suggestion that I am this mature hegemonic male, something which elicited a startled response: me, mature? Hopefully never! Following on from this flippant thought came a more reasoned – hopefully – response. I think that since the epiphany during my PhD, mentioned earlier in the chapter, I have sought to move away from binaries of human definitions. So to me there is no feminine or masculine side, no spectrum of either or both, no sliding scale upon which I can be positioned. Rather, there is simply 'me', in all my mixed-up complexity. Yes, some of my traits may be more definable as masculine or feminine, but these are only socially constructed labels, all of which can bind and restrict us and so are to be equally resisted and railed against.

Acknowledgements

When I had to give my inaugural professorial lecture at Otago (a fate I fought long and hard against), I reflected on how those people who in my academic life had been quasi mentors (my supervisors and heads of department) were exclusively male. Yet in writing this chapter, I think I have exclusively focused on women. None of them was ever in a management position relative to me but all have had a huge impact on my thinking. While I have no desire to do that lecture again, if I did I would have to seriously alter it. It may have been male mentors who have helped to shape me, but it is the female academics who have most shaped my thinking. Yet – and here comes the really soppy bit, so please look away all hegemonic males lest the sensitivities you do not have are offended – even the influence of these academic women pales into insignificance compared to the two women who have been most central to my life as a researcher, academic and person, and there really is no difference between those three. I would simply be nothing without my wife; I owe almost everything I am to her. I say almost all, because the rest is the responsibility of Snuffie (as well as Gypsy and Ebony who followed on from Snuffie), who like my wife changed my life irrevocably and in ways that I could never have guessed at and am still trying to understand. This hegemonic, individualistic male is simply nothing without them. In one last show of the hegemonic male within me, if you do not believe me on this issue or question the authenticity with which I make the claim I simply do not care, for on this one issue I know I am right; there is no doubt there in my mind.

Notes

(1) The fact that conceiving and/or having children in your family is not the sole domain of heterosexuals is of course a source of irritation to some misguided individuals who persist in seeing having and raising children as an exclusive domain of heterosexuals.
(2) If you do not know who Dog (as opposed to 'dog') is, then I urge you to cuddle up on the couch and delve into *Footrot Flats*, the creation of the great and very much missed Murray Ball.

References

Attwood, F. and Smith, C. (2013) Leisure sex: More sex! Better sex! Sex is fucking brilliant! Sex, sex, sex, SEX. In T. Blackshaw (ed.) *Routledge Handbook of Leisure Studies* (pp. 325–336). London: Routledge.

Berdychevsky, L. and Gibson, H. (2015) Women's sexual sensation seeking and risk taking in leisure travel. *Journal of Leisure Research* 47 (5), 621–646.

Berdychevsky, L. and Nimrod, G. (2017) Sex as leisure in later life: A netnographic approach. *Leisure Sciences* 39 (3), 224–243.

Bullough, V. and Bullough, B. (1977) *Sin, Sickness and Sanity: A History of Sexual Attitudes*. New York: Garland.

Carr, N. (2011) *Children's and Families' Holiday Experiences*. Abingdon: Routledge.

Carr, N. (2014) *Dogs in the Leisure Experience*. Wallingford: CABI.

Carr, N. (2016) Sex in tourism: Reflections from a dark corner of tourism studies. *Tourism Recreation Research* 41 (2), 188–198.

Munar, A.M. (2018) Researching in a men's paradise: The emotional negotiations of drunken tourism fieldwork. In B.A. Porter and H.A. Schänzel (eds) *Femininities in the Field: Tourism and Transdisciplinary Research* (pp. 170–184). Bristol: Channel View Publications.

Reichert, T. (2003) *The Erotic History of Advertising*. New York: Prometheus Books.

Schänzel, H. and Carr, N. (2016) Introduction: Children, families and leisure. In H. Schänzel and N. Carr (eds) *Children, Families and Leisure* (pp. 1–8). Abingdon: Routledge.

Smith, S. (1993) Bounding the borders: Claiming space and making place in rural Scotland. *Transactions of the Institute of British Geographers* 18 (3), 291–308.

3 Exploring the Expression of the Masculine in Adventure Activities: A Personal Reflection

Mark B. Orams

Writing this chapter has been a challenge. It has been so because I have never thought about my masculinity and its influence on my role as a researcher. In many ways this is surprising because we are taught to consider our potential biases when we approach research and either to seek to minimise them or to be explicit about them (Roulston & Shelton, 2015). I accept that I have a bias; actually I have many biases, and one of these is in my 'male-ness' and how this affects the way I view the world. It is interesting to me that I have not considered this before, ever, in any context. So, the purpose of this chapter is for me to share my reflections and learnings in considering my masculinity and how it affects my role as a researcher, particularly in the context of research into adventure. How do I do this, given I have not thought about it before? The logical place is to start at the beginning.

My Early Learning of What It Means to Be Masculine

In my childhood in provincial New Zealand, I grew, learned and understood that I was a boy. It was as simple as that. Being a boy was, to me, self-evident, unquestioned and unquestionable. It was reinforced in a multitude of ways: in clothing, hair-style, language, interactions with others, expectations and anatomy. It also contrasted with the 'other' gender – girls. I had two younger brothers and no sisters and my mother was not particularly 'feminine' (recognising that this, too, is a constructed and loaded term) and, as a consequence, I had little experience of girls. I thought of them as different, mysterious, mostly attractive and frequently confusing. I was taught to treat them differently. Physical interaction with other boys through play, sport and friendship was based on testing strength, both physical and emotional. Sometimes it was aggressive, and encouraged to be so. Fighting, wrestling and asserting yourself physically with other boys was expected. However, with girls this was discouraged. They were not considered physically as strong and we were not to engage with them in the same way as with boys. This was, and likely still is, a common differentiation for school-aged children (Klomsten et al., 2004).

The education system and society in 1960s and 1970s New Zealand reinforced these gender differences. For example, at my school (and at home) the punishment for transgressions (perceived or otherwise) for boys was physical (Maxwell, 1995). Boys were strapped in front of the class. That is, the teacher used a long leather strap to whip boys across the hands, legs or buttocks, repeated usually six times and with enough strength to raise welts and cause pain. As a boy, you were expected to endure this pain and show as little reaction as possible. If you were staunch through such an event, this meant you were brave and tough, which were highly revered attributes for a boy. Interestingly, if the teacher was a woman, a male teacher was called in to administer this punishment, further reinforcing the idea that physical aggression and interaction was a male behaviour. School policy did not allow girls to be strapped or physically punished. Actually, I cannot remember any punishment being dealt to girls during my early school years except a verbal 'telling off' – usually for talking in class.

My learned masculinity was, therefore, about strength, toughness, and controlling emotions so as not to show weakness. It was also about differentiating myself from the feminine, which was seen as soft, smooth, emotional, weak and non-physical (Gilbert et al., 1978).

In general, I avoided interacting much with girls as a child. I can remember being taught (although I am not sure by whom) never to hit a girl, and developing a sense that my role as a male was to protect girls, particularly from physical threats and danger, such as from other males. I also learned to be polite and courteous to girls and that this was part of being a boy and, eventually, a man. Examples were to alter my language so I did not use profanity or overly critical or belittling terms, whereas this was commonplace in communicating with other boys.

During my teenage years I was also taught to open car doors for girls/women, to offer up my seat for females and to allow them to enter through doors before me. This was considered to be polite and courteous. The general message was that females were different and that they were to be treated differently and this was expected from a man, especially a 'gentleman'.

I remember vividly when I was first confronted regarding this. I was an undergraduate at university in New Zealand and I went to open a car door for a female friend. She strongly objected to this and castigated me for my sexist behaviour. She then, and over subsequent conversations, educated me about the inherent sexism in such actions on my part. She argued that my pretence of politeness and courtesy was unacceptable because I extended such actions only to the female gender. Since this time, I have become more aware and much more careful in my use of language (much of our language is gender specific and reinforces gender stereotypes) and in my behaviour. However, this 'learning' has also been quite confusing because many females enjoy the 'old-fashioned' ways of courtesy towards women and are complimentary of such behaviour. Thus, my journey and approach of being cautious and guarded with regard to dealing with gender began, and still continues today.

What does this have to do with masculinity? It outlines my early understanding of gender and gender roles and the characteristics that are typically attributed to males, those being masculine, and those generally attributed to females, i.e. feminine. A further important reflection is that these understandings of masculine characteristics and feminine characteristics were reinforced by the generally disparaging attitude towards masculine attributes seen in females and, especially, the attitude towards feminine attributes seen in males, a phenomenon reported by Barnes (2011). This became particularly acute during my adolescence and early adulthood. Terms such as 'butch' and 'bloke-ish' were used to refer to females who had body hair (especially facial hair), who were of a solid or muscular build and who appeared more male. Conversely, males who expressed feminine characteristics were widely labelled as 'poofters', 'homos', 'queens', 'nancy-boys' and a variety of other derogatory terms associated with assumed sexual acts of male homosexuals.

Overall, then, the idea of masculinity that I developed growing up was the predominant one of the time and based on a simplistic bivariate understanding of gender, and a stereotypical set of attributes that were associated with being male (Young, 2001). Furthermore, feminine characteristics seen in males were viewed negatively (at least by other males) and masculinity was an admired set of attributes in males, but conversely not in females.

My Masculinity and Research

The personal context outlined above is important when considering how I both view and live as a male, as a researcher, as a tourist and in my recreational activities. I have a strong preference for recreational activities that involve action, competition, challenge, risk, adventure and nature-based settings. These activities allow, or require, the expression of what traditionally have been thought of as masculine attributes such as strength, courage, determination, endurance, aggression and toughness. Furthermore, I admire these characteristics in others and have 'heroes' (another strongly masculine term) who show these attributes in dramatic and obvious ways, and I aspire to be more like them. What is interesting is that while there is a wealth of literature exploring masculinity, femininity and gender in the context of organised and competitive sport (e.g.

Dashper, 2012; Harris & Clayton, 2002; Klomsten *et al.*, 2004; Messner, 1990; Pearson, 1982; Wellard, 2002; Whannel, 2005), there is much less in the adventure recreation area, and much of this work has explored the experiences of females as a minority gender within these communities (see, for example, Humberstone, 2000).

My experiences as both a participant and as a researcher in nature-based adventure recreation participation is that there is a clear male gender dominance. Adventure activities such as mountain climbing, surfing, sailing, sky-diving, para-sailing, base-jumping and so on have a high predominance of males (Humberstone, 2000; Bentley *et al.*, 2007; Humberstone, 2000). For many adventure activities, females are not only the small minority, they are a rarity. Interestingly, those females who are involved in such activities are viewed by male participants in two contrasting ways. They are often joked about and referred to as possessing masculine characteristics (for example, by jokes about her having 'balls' or being a 'she-man' or some other disparagement). Conversely, there is often respect and admiration for the attributes being shown in these female adventurers. There is, therefore, a certain 'bloke-ism' in these activities where traits, characteristics, behaviour and language are strongly aligned with the masculine. As a consequence, I feel very at home within these adventure recreation communities and have an innate sense of belonging and inclusion.

From a researcher's perspective, this provides ready access to these communities to collect data. It is easy to build a rapport and have credibility when I am a member of the predominant group and this is helpful when seeking opportunities to engage with potential study participants. Interestingly, when speaking with my wife about this she immediately emphasises how much easier it is for me to gain acceptance within these groups (such as surfers, sailors, scuba-divers, windsurfers) than it is for her. She often feels like an 'outsider' and an accessory to me in such scenarios. She also observes some changes in my behaviour when I am in the company of other male adventurers, changes which occur completely unbeknownst to me. For example, she observes that I tend to speak in a deeper voice, use profanity and colloquialisms more frequently, use more crass and disparaging humour, drink more alcohol and tell 'war stories' about my adventures and that I tend to ignore her. In considering her observations, sobering as they are, I can accept that subconsciously I modify my behaviour in these circumstances, in order to gain more ready acceptance into the group and to establish my credibility as a masculine male.

So, my masculinity does shape everything to do with my research, in the topics I choose to explore, the opportunities I have to gain access to social groups that are male dominated and even the behaviour I express in the company of other male adventurers.

Considering Masculinity and its Predominance in Adventure

The expression of the masculine in adventure-based activities also manifests itself in the language used. Actions that are seen as courageous are ascribed masculine attributes when being described to others. Examples include: 'that guy has big balls' (or *cajones* in Spanish); 'ballsy stuff!'; and I have heard a courageous female referred as 'she has testicles, that girl!'. Conversely, those who fail at or pull out of difficult or courageous acts are referred to disparagingly by ascribing feminine characteristics. Examples include: 'don't be such a big girl's blouse'; 'are you having your period?'; or simply by changing the use of the masculine to the feminine pronoun when referring to the individual (male), as in 'oh, she's decided not to do it' or 'here she comes, she's completely wussed out on that one'.

These observations of my own personal experiences are not unique. Media portrayals of the masculine are strong in depictions of nature-based adventure activities. These reflect societal views and also shape them. In the promotion of adventure-related products and experiences, strongly masculine imagery and language are used. Advertisements for fishing boats, surfboards, personal watercraft (jet-skis) and mountain bikes typically feature males actively involved in the activity and, if they are included in the imagery at all, attractive younger females in passive roles in admiring, submissive or sexually suggestive poses.

A typical example is the international surfing brand Billabong's online web-store front page, where the men's product section featured an image of a male surfer in action while the women's section used an image of a young, slim woman lying on a beach in a bikini with her back arched in a seductive pose. The reaction from Karen Banter (see https://you.women2.com/f-ck-you-billabong-seriously-f-ck-you-84995f3d7946) to this approach hit hard, right between the eyes (or, if you like, was a well-directed, figurative kick in the balls) when she stated in an opinion piece entitled 'Fuck you Billabong, seriously, fuck you':

> This is what you have to offer us. This is how you, as a company, see women, and it is also *what you are trying to sell us about ourselves*. Billabong makes products for men, who go out there like badasses and catch awesome waves, and also for women, who basically just lie around uncomfortably, waiting to be looked at and desired …
>
> Surfing is still such a Boys Club. Nine times out of ten, the make-up of people in the water is 90% male. Easily. Half the time, I'm the only female out there. Why continue to propagate this idea that *even according to a surf apparel company*, a woman's place is on the beach, not in the lineup? Do you not realise how damaging this is?

Even as a male, I see this kind of stereotyping everywhere. I know it's not right and I know it perpetuates stereotypes and encourages misogyny and sexism. As a consequence, I feel some guilt for my maleness (even though

I am not personally responsible for creating such imagery or communications). The more I look, the more I find such male domination all over the place. For example, extreme adventure stories that feature in the news media are dominated by men. Since their inception, movies have depicted men in lead roles where they show courage, face fear and overcome challenges and life-threatening situations (Lehman, 2013; MacKinnon, 2003: 116). Superheroes in comic books are male dominated (with a few exceptions) and 'war heroes' who are recognised and awarded for bravery are almost all men (Taylor, 2007).

So, my own personal experiences of the domination of the masculine in nature-based adventure are widely portrayed in society. Stereotypes abound of positivity and admiration for the masculine and the heterosexual in adventure. Conversely, while more subtle, the messaging is widespread that females do not belong, and that feminine characteristics expressed in males are negative attributes.

In thinking about this, it occurs to me how difficult it must be if one's natural tendency is towards the feminine when you are a male and vice versa, and even more so if one's birth gender is completely alien to yourself and you feel that you are not your biological gender – or not strongly feeling being either male or female. We have such a simplistic bivariate view of male and female and masculinity/femininity. Fortunately, this is changing. This journey to a more enlightened, inclusive and accepting view of the diversity of gender is currently underway, both for me personally and I think more widely. Like all significant societal changes, this journey is difficult and controversial. It challenges existing long-held beliefs and tests people's previously held worldviews. Unfortunately, it brings out prejudices, bigotry, hatred and cruelty in some. Throughout human history such changes have been hard fought. Examples include the ending of slavery and institutionalised racism in many countries, the women's suffrage movement and rights to equality, decriminalising homosexuality and then legalising rights to same-sex marriage. What these experiences have shown is that the journey is long and difficult and produces many casualties.

From a researcher's perspective, it now occurs to me how much more difficult it must be for women and more feminine men, or any LGBTQ person, to conduct research into adventure recreation and, especially, into the experiences of men. Perhaps this is understandable, because for any 'outsider' to gain access to hegemonies is difficult. However, outsiders often see things that an insider cannot, as the conversation with my wife about my own behaviour revealed!

What about Femininity?

I am not suggesting that an exploration of masculinity in nature-based adventure activities is on the same level as the above-mentioned significant and important human rights battles. However, engaging in nature-based adventure and the opportunities for growth, health and enlightenment should not be the exclusive reserve of the masculine. The expression of the feminine, irrespective of one's gender, can add to the diversity of the human experience when interacting with nature. To feel deeply the full range of human emotions and attributes must surely add to the opportunities such activities provide. What if we were more accepting of the expression of the feminine in such activities? Could this add something of value?

What Does This Mean for Research into Adventure?

These personal reflections, and the revelations that have resulted, may be of value for other researchers when they approach studies on adventure-based experiences. For any social science researcher, especially when using qualitative approaches such as phenomenology or grounded theory, the ability to connect with and be accepted by the study participants is important. In adventure activities, where masculine attributes such as courage, toughness, resilience, endurance and fortitude are an inherent part of the risks involved, establishing respect and credibility as a researcher is critical for access and acceptance and for meaningful sharing to occur.

The neo-tribes that exist in such adventure activities tend to be fiercely independent and judgemental. You are either accepted as one of them or you are not. This acceptance is more likely when the researcher is known to and respected by the participants, in order for them to accept the researcher as 'one of them' (as opposed to an outsider) and feel comfortable sharing their experiences, using their own language terms and sharing their stories knowing they will be understood.

Surfers, for example, are a strong 'neo-tribe' with their own cultural norms and expectations. A non-surfer asking questions about their experiences in participating in their sport will typically be treated with caution (or even disrespect) and they may even struggle to understand the language used to describe those experiences (e.g. 'This complete kook dropped in on me when I was getting totally shacked and I got sucked over the falls and then clubbed by the lip. It was such a sick barrel and I just got worked. It wrecked my session and I was dragging the rest of the day.').

This is not to say that it is impossible for a female researcher to establish such a connection, but I think it would be more difficult. My research into the effects of ocean-based adventure sports on participants has been dominated by male participants because the sports I have investigated (e.g. surfing, sailing, ocean-swimming, paddling) are all strongly male dominated. I think it would be challenging for a female researcher to gain access (to gain agreement to be interviewed in the first place) and for her to have participants share openly and honestly with her. In my experience, there is a 'filter' that is used in such scenarios and responses are subconsciously more guarded because of a female presence.

To give a specific example, the following data excerpts are taken from a one-on-one semi-structured interview with a male experienced offshore sailor. He shared this about the views of ocean sailors on anyone aspiring to join their community:

> They (offshore-sailors) are brutally honest … He would get cut no slack, absolutely no slack. He'd have a decision to make. He would either handle it, toughen up, or not put himself through it.

Even more dramatically, the judgement made about a man who made a last-minute decision to get off the yacht and not go to sea:

> For a guy to step onto the yacht and throw the dock-line onto the beach takes guts. I've seen guys jump over the side and swim away. I've seen that. They just can't do it. They thought they could do it, but when it comes to it, they can't. … The guy who jumped over the side. That decision defined him, and he's a fuckin' idiot. He's a lemon in my mind. A gutless lemon who let down his team. He's not gonna get the male love from this Kiwi I can tell you.

Conversely, the views about the crew-mates who had the courage to take on the adventure are ones of admiration and respect:

> Look into the eyes of a man who's just sailed a leg of an ocean race, you're looking at a man who's incredibly confident, he's incredibly peaceful, and he's tired. But, he's a man who's achieved something way more than a salary. It's a very beautiful thing to see, a man that's experienced it.

It is notable that all of the responses always use the masculine terms 'he' and 'man'. This is understandable because all of the respondent's experiences in the activity have been exclusively with other men. I am certain that if this same interview had been conducted by a female interviewer the responses would have been 'filtered' and the responses less direct, less judgemental and less honest.

So, for a female to work as a researcher in the area of exploring the experiences of male adventurers, she would need to work harder to establish her credibility and to be accepted. Often this credibility comes from past achievements and experience. Reputation is also important. There are, for example, a number of females who have completed around-the-world yacht races and who have survived extreme conditions and challenges. Such achievements do engender respect from men and, as a consequence, provide access to sharing with the 'filter' less in effect.

Interestingly, in my interviews with men where I explore what their long-term involvement with adventure sports means to them, responses are sometimes quite emotional. On numerous occasions interviewees have 'teared up' when reflecting on the importance of the sport to them and their lives. They feel deeply about these things and this is evidenced in this reaction, a characteristic more typically seen as feminine. Perhaps, therefore, the 'feminine' does come through and add value as I advocated in the previous section. When I have questioned the men about their emotional reaction they have all expressed surprise that this happened to them. They did not expect to become emotional during their responses and in sharing their reflections and they were quite taken aback by it. I wonder if such vulnerability would have made it through the 'filter' if they were being interviewed by a female?

So, What Does This All Mean?

In sharing these reflections in this chapter I find myself feeling a little concerned and vulnerable. It is almost as if I am betraying what some have termed the 'bro-code' – that being a kind of unwritten expectation that men do not share the key aspects of what it means to be a man. There is truth here, though (or at least 'my truth'), and sharing truth can sometimes be confronting and difficult. In writing this chapter and, especially, in the self-examination that has been needed to write it, I find that my masculinity affects everything I think and all that I do. It shapes who I admire, who I befriend, how I view my own worth and success, who I am attracted to, how I treat my children, the language I use, the clothes I wear, the things I aspire to. Inevitably, therefore, my masculinity has a huge influence on my role as a researcher. It must, because it is unavoidable. It is a bias. It cannot not be. I am who I am, and I'm glad I'm a man (to borrow a part of a line from 'Lola', The Kinks' classic song exploring transgender sexual attraction). But, what I am learning is that being a man also means I can accept that femininity is part of my being and that allowing myself to experience that does not lessen my manhood or diminish my sense of who I am. It adds to it.

Final Thoughts

Thinking about and writing this chapter has been a journey in self-examination, self-discovery and learning. Such contemplation and realisations are uncomfortable because they have caused me to question myself and my approaches to life and to the other, that is, people who are different from myself. On the surface, gender seems so simple: you are either male or female and, similarly, a male is masculine and a female is feminine. The reality, however, is far more complex and nuanced than this, something that feminist researchers have been saying for decades. Perhaps we men are finally listening and becoming more enlightened as a result.

References

Barnes, C. (2011) A discourse of disparagement. *Young* 19 (1), 5–23.
Bentley, T.A., Page, S.J. and Macky, K.A. (2007) Adventure tourism and adventure sports injury: The New Zealand experience. *Applied Ergonomics* 38 (6), 791–796.

Dashper, K. (2012) 'Dressage is full of queens!': Masculinity, sexuality and equestrian sport. *Sociology* 46 (6), 1109–1124.

Gilbert, L.A., Deutsch, C.J. and Strahan, R.F. (1978) Feminine and masculine dimensions of the typical, desirable, and ideal woman and man. *Sex Roles* 4 (5), 767–778.

Harris, J. and Clayton, B. (2002) Femininity, masculinity, physicality and the English tabloid press: The case of Anna Kournikova. *International Review for the Sociology of Sport* 37 (3–4), 397–413.

Humberstone, B. (2000) The 'outdoor industry' as social and educational phenomena: Gender and outdoor adventure/education. *Journal of Adventure Education and Outdoor Learning* 1 (1), 21–35.

Klomsten, A.T., Skaalvik, E.M. and Espnes, G.A. (2004) Physical self-concept and sports: Do gender differences still exist? *Sex Roles* 50 (1–2), 119–127.

Lehman, P. (2013) *Masculinity: Bodies, Movies, Culture*. London: Routledge.

MacKinnon, K. (2003) *Representing Men: Maleness and Masculinity in the Media*. London: Arnold.

Maxwell, G. (1995) Physical punishment in the home in New Zealand. *Australian Journal of Social Issues* 30 (3), 291–309.

Messner, M.A. (1990) Men studying masculinity: Some epistemological issues in sport sociology. *Sociology of Sport Journal* 7 (2), 136–153.

Pearson, K. (1982) Conflict, stereotypes and masculinity in Australian and New Zealand surfing. *Australian and New Zealand Journal of Sociology* 18 (2), 117–135.

Roulston, K. and Shelton, S.A. (2015) Reconceptualizing bias in teaching qualitative research methods. *Qualitative Inquiry* 21 (4), 332–342.

Taylor, A. (2007) 'He's gotta be strong, and he's gotta be fast, and he's gotta be larger than life': Investigating the engendered superhero body. *Journal of Popular Culture* 40 (2), 344–360.

Wellard, I. (2002) Men, sport, body performance and the maintenance of 'exclusive masculinity'. *Leisure Studies* 21 (3–4), 235–247.

Whannel, G. (2005) *Media Sport Stars: Masculinities and Moralities*. New York: Routledge.

Young, J.P. (2001) Displaying practices of masculinity: Critical literacy and social contexts. *Journal of Adolescent and Adult Literacy* 45 (1), 4–14.

4 Meditations on Masculinity: Encounters in Salty Research Spaces

Jacques D. Mahler-Coetzee

Background

I do not remember thinking about my masculinity as a child. I just 'was'. I think that might be the most natural state of being, even, I would hazard, as an adult. I recall a Buddhist sentiment along the lines that 'we all start out perfect and spend the rest of our lives trying to get back to that state'.

Nevertheless, what follows are several primary meditations on the notion of masculinity as a material factor personally experienced in my academic research spaces. These constitute an ever-emerging, always incomplete, developing narrative of this author as a male researcher adventuring at the nexus of several discourses in the global South.

My primary discipline for academic purposes is the Law. I do not hold myself out as a tourism or gender researcher, despite having considered the former in the African context through the lens of surfing (Mahler-Coetzee, 2018) and the latter in jurisprudential writing on legal status (Mahler-Coetzee *et al.*, 2017: 83–90). However, I see legal systems as anthropological phenomena, telling us as much about the human condition/ourselves as they inform us about rules with which we seek to gloss our original organic state with some momentary semblance of certainty. As such I find myself in applied spaces, researching phenomena and issues associated with, for example, coastal development, tourism governance, economics, marine spatial planning, higher education, pedagogy, recreation, heritage, ethics, piracy, brand semiotics and surfing, among others, many of which impact and speak directly to themes and concerns of the tourism field. In these, I am drawn to non-positivist, grounded approaches and humanist points of entry.

Whether we define ourselves by tourism or not, we also operate in the incrementally larger fields, of research, higher education and academia in general. My meditations also arise within a unique political

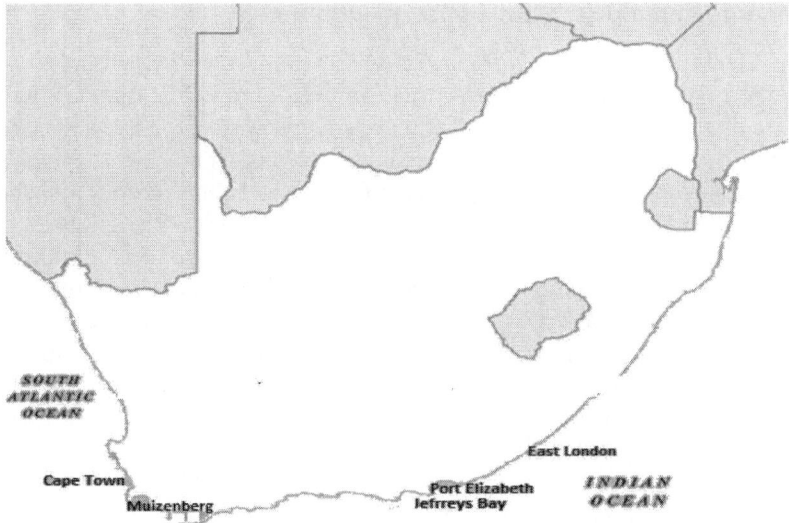

Figure 4.1 South Africa map (Distance East London to Cape Town = +/−1027 km/ 638 miles

Source: Created with MS Paint using line map from http://www.mapsopensource.com/south-africa-outline-map-black-and-white.html © J. Mahler-Coetzee.

jurisdiction, the history and challenges of which undoubtedly inform and influence my perceptions. My reflections on experiences in research-related fieldwork reference a particular geographical context and, in some instances, are materially defined by particular locations and the distance between them. The map in Figure 4.1 provides a shared frame of reference in this regard.

The three 'case study' meditations below are a record of some key waypoints, charting my course of discovery of the world and of self. 'In my beginning is my end' (Eliot, 1936a: 196), so I start with an encounter with my masculinity as a student at university, through encounters with myself 'looking at myself' in research collaborations, and finally encounters in researching (in this case, women) surfers.

I will try 'to be honest, to do no harm, and to give each man his due'[1] (Moyle, 1913: Proemium).

Requiring to Be Admired

A benefit of higher education is often what one learns outside the lecture hall. An initial, modest, positively disruptive encounter with early stereotypical perceptions of masculinity and 'otherness' set the tone, dare I say, for my future more enlightened self.

Inherited prejudice, constructive enlightenment

I feel that every adventure begins with a map, or with a willing companion or, on rare and magical occasions, with both. The map (a small part of a larger African continental geography) was of the University of Cape Town campus, and she, the companion, was Mary.[2] I was trying to locate my venue for an undergraduate English literature tutorial and she showed me the way. It was raining; we shared my black umbrella. Sean, a male student, entered the venue under a pink umbrella. I recall making some disparaging remark to Mary regarding Sean's lack of manliness based on the, at that time, prevailing 'pink is for girls' stereotype.

I do not quite know why I said that – perhaps to impress upon the female I liked how masculine I was. This was the funny way my naïve and uncritical boy-man's mind worked at that time. This was, some 30 years ago, in the pre-democratic South Africa of the late 1980s, where I grew up in the intangible cage of a pervasive conservative masculine rhetoric which I recall was marked by, among others, xenophobia, homophobia and disdain for 'women's lib'.

I was shocked by Mary's swift rebuke. She looked at me in entirely the way I did not want to be looked at by one I profoundly admired and whose admiration I sought. She expressed in no uncertain terms that my comment meant nothing as far as the sexuality of Sean was concerned, but rather suggested some uncertainty as to my own sexual security. I was discombobulated ... for a while. I returned to her and university the next day forever altered, enlightened. Mary was astute enough not to dismiss me, but to constructively chastise, to unburden me of oppressive narratives, to educate me.

Reflecting on this formative disruptive experience some three decades later, I realise that:

> ... the male gender role itself is kind of conceptualized as a precarious status. Manhood is something that is hard to earn and easy to lose ... (Bosson, cited in Vendantam *et al.*, 2018)

and that the experience was, and is still, essential to my continued personal negotiation and navigation of an enlightened masculine identity.

Collaboration and Competition

In relation to academic collaboration (particularly in the context of male scientists employing females on their teams), the term 'Pickering's harem' or the 'harem effect' has, despite its tabloid-like nature, been a historical phenomenon associated with some surprising, if isolated, progressive thinking (Geiling, 2013). Previously, among other factors, women were cheaper to hire than their male counterparts, they provided equivalent or better competencies (Geiling, 2013; Rossiter, 1974, 1980 generally; Elkin, 1901, cited in Bly, n.d.) and afforded less potential professional competition. While these pragmatically mundane underpinnings of the phenomenon are informative of the research milieu of a century ago, they also have persisting resonance with my experience of the competitive aspects of current academic research dynamics.

Indeed, when working with male collaborators, the question of masculine competitiveness is ever-present on the periphery. When, for example, a male supervisor notes how 'cool' one is (with reference to one's dress/style/way of interacting with other, especially female, colleagues), this is not necessarily an innocent remark; it may be one of personal comparison and possibly, in a research context, unhealthy, personal competition. By contrast, when a female collaborator has said something similar, I have appreciated such remarks purely as a compliment and a sign that the collaboration is working well.

My experience of collaborating on, and eventually leading, an institutional team on a national pedagogic project resonates. However, unlike the classic 'harem' noted above, my female team colleague co-opted me. Her rationale was that she, being at that time relatively inexperienced in academia and project research, required my experience (Jawitz & Perez, 2014: 57). I noticed almost from the outset, in the company of others, especially the male participants from other institutions, a protective and possibly possessive feeling in me. I cannot recall ever expressing this directly, but I did indeed feel an internal tension in this regard. This was at least in part not a result of a trait of masculinity, but rather a functional concern in protecting the integrity and secrecy of what was my substantial intellectual capital input in our local project. I felt that my inexperienced colleague might unwittingly be compromised in a context where I had noted such unaccredited purloining before. However, I experienced this 'us against the world' interpersonal dynamic to be understood, welcomed and reciprocated by my colleague, whose verbatim sentiments were captured by the meta-researchers as follows:

> ... where previously the lecturer [my colleague] had found that the ethos of the department meant she would have been likely to work mostly independently, the project had 'changed that for me because of working with Jacques, where we met often and we discuss ideas and we get excited ... ja, it's been a learning curve'. (Jawitz & Perez, 2014: 61)

This positive, dare I say joyful, collegiate relationship, being obvious to non-project male colleagues in the ordinary faculty context, led to a rather personal 'pushback' against the project and our team duo. We perceived this to be composed of several vectors: (a) jealousy at my being able to spend exclusive time with the female colleague; (b) a form of functional envy at the 'privileges' afforded by the project, such as teaching relief (Jawitz & Perez, 2014: 64); and (c) the result of structural and personality politics. This composite of pressures, in part, led to the resignation of my teammate from her employment, and thus our fruitful, nuanced, creative, collaborative research relationship was broken.

Chasing the Shadows of Butterflies[3]

I was finding in my reading around my doctoral thesis (which concerns the governance and potential legal regulation of recreational spheres, among others, surf tourism and surfing in South Africa; Mahler-Coetzee, forthcoming) ideas and issues that begged the question of post-millennium women's experience of recreational surfing (see, for example, Wendt, 2015). Noting that 'Surfer voices are best heard at the beach rather than in texts' (Posel, as cited in Thompson, 2015: 3), I took a grounded approach, seeking to obtain first-hand accounts from surfing women, and to contrast these with what little existed at the theory and macro level in the writing of local academics – at the time, predominantly male authors. I saw this work as having the potential to provide additional perspectives and background for my own understanding of current issues in surfing and, of particular concern to my doctoral work, how surfing might be more inclusively governed and/or regulated in an informed, responsive way.

I thus sought to profile three iconic South African surfing women. The first (accomplished with reference to archival materials) was the late Heather Price, hailed as the first person, male or female, to be recorded as surfing 'standing up' in South Africa (International Surfing Association, 2014; ZIGZAG, 2014). I then undertook several in-person, unstructured conversations with two further iconic surfing women: one from the 1960s/1970s era, Cheron Kraak (still the leading personage in the South African surf industry and formerly head of Billabong SA; Hift, 2017; Jarvis, 2017); and the second with a millennial 'icon ingénue', Jennifer Bam, identified by ZIGZAG surf magazine as the first commercial female surfboard shaper in the country (Paterson & van Gysen, 2016).

Accessing women surfers: Challenges for a male researcher

Previously, before the initiative noted above, I had naïvely failed to consider my masculinity as a factor that might impact my research logistics. For example, at my local surf spot in the Eastern Cape a local 20-something female surfer was a semi-regular sight on the beach, and although we had surfed the same break and exchanged greetings for several years, I did not know her. In my enthusiasm, I made the error of not getting some of the local male surfers (who did know her) to give me an introduction, but approached her directly after a surf, to tell her about my proposed research and ask if she would consent to an interview, possibly recorded either in sound or video. The mention of recording/video immediately garnered a defensive, suspicious stance and attitude in both body language and tone. Despite providing her with my business card credentials, I never heard from her again. Research lesson learned.

I thus became acutely aware of the potential difficulties that might await me as I contemplated collecting data through formal interviews with Cheron and Jennifer, in being a male wishing to access females' perspectives for research purposes. I also came to know the importance of gatekeepers to this process – in my case, male surfers who would have to vet and vouch for me before I could gain access to these women. This is an interesting observation of a protective dynamic, a stereotypical attribute of the masculine, which in its application in this context did not appear to register as negative (for example, as possessiveness) in the eyes of the females concerned. The processes of access, such as the vetting by these gatekeepers, was similar despite the vastly different personal circumstances of the two individual women recruited for the research.

Case #1

In the case of Cheron Kraak, I had to physically present myself at her business offices in Jeffrey's Bay (see Figure 4.1) and reiterate my purpose and request for an interview to front of house/personal assistants. As she is sought after by the press and media, I had to distinguish my research interest and position from that of the media. My details were taken, but it was suggested that I should make contact with Yuri, an established male figure in the Jeffrey's Bay surf scene. I had previously seen but not interacted with Yuri at one of several public meetings, on whose committee I sat as the nominal academic representative of the local university stakeholder regarding the Jeffrey's Bay Community World Surf Reserve application initiative. I did not gain access to Cheron Kraak on this occasion, nor was I able to link up with Yuri; I had to drive 300 km down the coast from East London again later to meet with her, at which meeting Yuri was also present. This initial meeting was not recorded; it morphed into a general conversation about a variety of topics, which I felt was essentially about establishing my own surfing credentials.

To clarify 'surfing credentials', being a surfer is not just about the act of surfing; it involves a complex composite of attributes, observances and expressions, mostly unwritten or informal, which are given meaning and recognition by other surfers. Factors that influence one's acceptance by another surfer, among others, include where one surfs, what one surfs, how long one has surfed, how one dresses and looks, how one talks about surfing, and a general surfing ethos. Establishing such surf credentials is essential for supporting naturalistic and ethnographic research which depends on 'matters of access and permission, establishing a reason for being there, developing a role and persona, identifying gatekeepers who facilitate entry and access' (Cohen *et al.*, 2007: 178). As with other fields of ethnographically grounded research, establishing authentic 'rapport, trust, sensitivity and discretion' with one's research participants is key to gaining effective access to and meaningful insight into their community, cultures and practices (Cohen *et al.*, 2007: 181; Welman *et al.*, 2005: 196).

An example of my own deeply felt authentic absorption of surf ethos and symbology in my appearance at the time was the length of my hair, worn long and/or in a ponytail, as well as a series of henna tattoos (Figure 4.2, first row). This seemed to be appreciated as one sign of my 'surf-worthiness', appearing to mitigate a possible adverse inference attributed to my legal and academic occupation, and going some way to disarm the gatekeepers of access to my surf community research participants. Indeed, it was not long after that meeting that I received a phone call to confirm that we might arrange to have a recorded telephone interview about women in surfing in South Africa.

Case #2

In the case of Jennifer Bam, a recent fine arts graduate from a local university, now living in a rural area (with a cow and a Jack Russell!), the road to an eventual telephone interview (Bam, 2016) took some bizarre geographic turns. Having become aware of her through the lead article (Paterson & van Gysen, 2016) in *ZIGZAG* surf magazine, I contacted the male author of that piece to see how I might be brought into direct contact with Jennifer. After making clear my research intentions and that I would not be competing with his own work/sphere, he passed me on to Cobus Joubert of WaWa Wooden Surfboards™, who had initially recognised and supported Jennifer's talent and through whose enterprise Jennifer sold her boards. After a lengthy telephone conversation, he agreed to pass my details on to Jennifer, and to leave it to her to respond if she felt like it.

In the interim, I presented myself in person to Cobus at his surf store, some 1067 km away (see Figure 4.1) in the Cape. Apart from 'putting a

Figure 4.2 Establishing surf credentials collage (all images © J. Mahler-Coetzee)

face to the voice' and confirming my bona fides regarding my request for access to Jennifer, I felt we found common ethical ground. Our discussions moved across a variety of surf-related topics, with a special emphasis on the original ethos of surfing and how the same informed both Cobus' 'tree, sea, soul™' (see http://wawawave.com) sustainable surfboard-making enterprise, as well as my own appreciation of 'old school' style and modern green sensibilities demonstrated, among others, in what I rode, an agave triple-stringer retro fish.[4] The understandings established at this initial research-inspired meeting have endured beyond that research occasion and happily count among my recurring positive interactions on my regular visits to Cape Town.

I then waited for some time before Jennifer did indeed contact me. When she did, it was clear that I had to allay fears as to my 'intentions'. It helped tremendously that she had studied at Rhodes University, where I had been previously employed and where my younger sister had studied. Familiar points of reference, as well as being a surfer, cemented trust.

I realised that there was a back story to this female participant's initial wariness of my motives. In the subsequent interview, Jennifer related her disappointment at how a seemingly friendly offer of a surf by a local male while she was travelling in California morphed into discomfort when it became apparent that his motives were not purely altruistic:

> But I found this little cool point bay that was working and there was like three guys in the water and the one guy just got out and said, 'Oh wow, you look so cool', you know like having a good time. 'You know what, take my board, take my wet suit, it's almost high tide like get what you can.' … I was so stoked. … Ja, it was really cool. The guy was really sweet and then he ending up taking me to look at this cove which is really cool; I wouldn't have found it otherwise. I mean I was just walking on foot … But then it always ends up in the same … maybe it's just my dynamic that I deal with 'it's me in this world' but then, you know, the guy was start taking a fancy to you and like, okay, got to go home now. (Bain, 2016, 21)

Being presented thus with a basis for Jennifer's initial general wariness of even a seemingly bona fide male, I started to comprehend the underpinnings of a possibly general gatekeeping phenomenon. While this type of initially negative female generalisation/reaction is difficult not to take personally by a male researcher, it must be understood to reflect the realities of the world in which these participants exist, specifically a 'fear' of unwanted male attention/access. Indeed, I have not experienced a comparable emotion when being a male research participant for female researchers. It is thus clear to me that as a male researcher of female participants, such free-floating, seemingly commonly shared, collective 'fear' is a factor of one's masculinity that must be acknowledged and which must be planned for when venturing into the field.

Indeed, there are hopeful hints that the basis for the perceived barrier noted above is being eroded elsewhere, such as in the example below, recognising 'male allies' in the surfing field:

> In a Facebook post, the league announced that women surfers will receive equal pay at all events from 2019. In the world of surfing – a sport and culture long dominated by men – this is a monumental development. A range of issues, including women's activism, international sport policy change, female leadership and male allies, have led to this decision. The factors might be unique to surfing but they illustrate the complex ways in which significant gender changes come about in some sports. (Thorpe & Wheaton, 2018)

Fear, mediation and gatekeeping

It is apparent that being a male, no matter how friendly or how backed up by research credentials, causes initial suspicion of 'ulterior non-academic personal/romantic motives' in female participants. What is interesting is the male gatekeeping structure/system employed by or surrounding female surfers to mediate this. The positive protective role of the classic masculine traits appears usefully employed to counter another negatively perceived/unwanted trait associated with being male. A male researcher, it appears from this, is inherently seen as a potential sexual threat, a stereotypical male first and, only after vetting (ironically, by other males), a non-threatening researcher second. This hurdle must ultimately be viewed pragmatically by the male researcher, reflecting on it as an abiding element of the female experience across fields.

Fear too, original sin and the chilling effect

Given the existence of the fear above, and a larger gender discourse built on narratives much like Jennifer's, for the unarmed and disarmed male researcher acting in good faith another apprehension arises: a fear of rejection, and a sense of being inherently suspect. Consequently, I was reluctant to engage with topics touching on gender, even in a not-directly-related discipline such as tourism. Indeed, before my engagement with female surfers noted above, and before this very chapter, gender issues felt an emotionally confusing space too closely tied in with personal identity to risk a seemingly inevitable backlash – in essence a battle to be avoided. I am not sure that this challenge, a type of self-censoring phenomenon, is experienced in this way by female researchers.

My sense of this acutely felt barrier to my exploration is disappointing on many levels – because we lose the potential of another valid voice in spaces where gender is at issue, but primarily also on principle because I would like to be able to 'write what I like' (Biko, as cited in Stubbs, 1987:

title page). Actualising this realisation, this chapter is, to some extent, my part in personally overcoming such perceived limitations.

Conclusion

The meditations above suggest some observation lessons about the impact of 'masculinity' in the journey through some of my own research experiences. They are not, and will never be, complete. My goal is also not necessarily agreement, but stimulation. It may be that 'You are you and I am I, and if by chance we find each other, it's beautiful. If not, it can't be helped' (Perls, as cited in Dolliver, 1981).

Along the way, I have found that the perceived stereotypical encultured traits associated with generic masculinity, where these are unproductive, may be disrupted. I have exposed a potential logistical challenge for male researchers wishing to access and gain the trust of female research participants. In doing so, it has also become apparent that the experience of the female as a research participant may be fraught with concerns that generally might not have an equivalent in my own experience as a male research participant. Of interest is an observation that there are spaces in which masculine stereotypes are embraced/leveraged, such as in the gatekeeping function by male surfers for female surfers.

In contrast to generally creative and nurturing academic collaborations with females, I have experienced masculine competition compounded by structural hierarchies and power dynamics in academic institutional contexts, as an unproductive distraction from the research enterprise. Equally, the recognition of a type of apprehension or reticence, as an emotion felt by a male researcher seeking to engage with gender issues in academic discourse, is an interesting phenomenon I have had to interrogate in myself.

Looking Through Me I see You

My antidote has been the constructive facilitation of female allies. Through proactive engagement in this reflective exercise, consciously exploring masculinity through the lens of my research experiences, I have acquired a set of referents for entry into the gender discourse that I did not formerly or formally have. I can now begin to talk, and therefore exercise my mind, about and in relation to things that previously had no name for me. My reflections have caused me to 'arrive where [I] started and know the place for the first time' (Eliot, 1936b: 222). For this, I am extremely grateful.

The grand challenges we face as a human species do not make distinctions based on sex or gender; we are all infinitely disposable in the cosmic scheme of things. Yet, as individuals, we have great relevance and profound meaning for each other. Research is a privilege and those who research bear great responsibility. I remain dedicated even more now to improving my representation of my fellows (sisters and brothers), to holding up a mirror to them, in their service. Looking through me, I see you.

Notes

(1) Translated from the Latin, the founding precepts of my discipline, the Law: '*honeste vivere, alterum non laedere, suum cuique tribuerum*' (per Justinianus Imperator, in Moyle, 1913: Proemium).
(2) Pseudonyms have been used to protect the anonymity of real-world characters, appearing in my meditations that follow, who have not given express permission for disclosure of their identities.
(3) From Bam's (2016) transcript, describing the antics of her Jack Russell terrier while she was on the phone to me.
(4) A 'fish' is an older type of surfboard with a two-pointed 'fish'-tail.

References

Bam, J. (2016) Personal communication/interview transcript.
Bly, E. (n.d.) *The Harem Effect on Female Scientists in the Victorian Era*. See http://racingnelliebly.com/weirdscience/harem-effect/.
Cohen, L., Manion, L. and Morrison, K. (2007) *Research Methods in Education* (6th edn). New York: Routledge.
Dolliver, R.H. (1981) Reflections on Fritz Perl's Gestalt prayer. *Personnel and Guidance Journal* 59 (5), 311–313.
Eliot, T.S. (1936a) 'East Coker', Part II of *Four Quartets*. In *Collected Poems 1909–1962*. London: Faber & Faber.
Eliot, T.S. (1936b) 'Little Gidding', Part IV of *Four Quartets*. In *Collected Poems 1909–1962*. London: Faber & Faber.
Geiling, N. (2013) The women who mapped the universe and still couldn't get any respect. *Smithsonian Magazine*, 18 September. See https://www.smithsonianmag.com/history/the-women-who-mapped-the-universe-and-still-couldnt-get-any-respect-9287444/.
Hift, R. (2017) *Cheron Kraak: Originator of Billabong South Africa*. See http://www.crowsnest.co.za/content.asp?PageID=453.
International Surfing Association (2014) *Historic National Surfing Landmark to be established at Muizenberg Beach, South Africa*. See https://www.isasurf.org/historic-national-surfing-landmark-established-muizenberg-beach-south-africa/
Jarvis, C. (2017) *Cheron Kraak: The Matron Saint of South African Surfing*. See https://www.swellnet.com/news/swellnet-dispatch/2017/11/23/cheron-kraak-matron-saint-south-african-surfing.
Jawitz, J. and Perez, T. (2014) *Managing Teaching Development Grants: Experiences from the Large Classes Project*. Report to the National Symposium, Cape Town, April 2014. Cape Town: CHED and CHE.
Joubert, C. (2016–2020) WaWa wooden surfboards™. Personal communication.
Mahler-Coetzee, J. (2018) From fringe to core: Contemplating surfing's potential contribution to sustainable tourism development in South Africa. *Tourism in Marine Environments* 12 (3–4), 221–238.
Mahler-Coetzee, J. (forthcoming) The legal regulation of recreational surfing in South Africa (provisional title). LLD thesis, Nelson Mandela University. See https://lawofthesea.mandela.ac.za/Researchers/Doctoral-students/Mr-J-Mahler.

Mahler-Coetzee, J., Barratt, A. and Denson, R. (2017) Sex and gender – status and capacity. In A. Barratt (ed.), W. Domingo, W. Amien, R. Denson, J. Mahler-Coetzee, M. Olivier, F. Osman, H. Schoeman and P Singh *Law of Persons and the Family in South Africa* (2nd edn). Pinelands, South Africa: Pearson.

Moyle, J.B. (1913) *The Institutes of Justinian* (5th edn). Oxford: Clarendon Press.

Paterson, D. and Van Gysen, A. (2016) Catching up with the surfer behind the planer: Jennifer Bam. *Zigzag* 40 (5), 64–71.

Rossiter, M.W. (1974) Women scientists in the United States before 1920. *American Scientist* 62 (3), 312–323.

Rossiter, M.W. (1980) 'Women's work' in science: 1880–1910. *Isis* 71 (3), 381–398.

Stubbs, A. (ed.) (1987) *Steve Biko: I Write What I Like – A Selection of His Writings*. Oxford: Heinemann.

Thompson, G. (2015) Surfing, gender and politics: Identity and society in the history of South African surfing culture in the twentieth-century. PhD thesis, Stellenbosch University.

Thorpe, H. and Wheaton, B. (2018) *Women's Surfing Riding Wave towards Gender Equity*. See http://theconversation.com/womens-surfing-riding-wave-towards-gender-equity-103299.

Vendantam, S., Shah, P., Boyle, T. and Cohen, R. (2018) Man up! How a fear of appearing feminine restricts men and affects us all. *NPR*, 1 October. See https://www.npr.org/2018/10/01/653339162/-man-up-how-a-fear-of-appearing-feminine-restricts-men-and-affects-us-all.

Welman, C., Kruger, F. and Mitchell, B. (2005) *Research Methodology* (3rd edn). Cape Town: Oxford University Press.

Wendt, L.S. (2015) What is it like to be a surfer girl? A phenomenological exploration of woman's surfing. PhD thesis, Southern Cross University.

ZIGZAG (2014) The Berg – same as it ever was (Muizenberg to get historic national landmark). *ZIGZAG*, 4 March. See https://www.zigzag.co.za/featured/the-berg-same-as-it-ever-was-muizenberg-to-get-historic-national-landmark.

Part 2
Performing Heteronormative Masculinities

5 Performing and Negotiating Filipino Masculinities in the Field

Richard S. Aquino

Introduction

Gender has been argued as a performance that is usually confined within the prescriptions of one's culture, such as the roles and expectations surrounding heterosexuality (Jagger, 2008). This notion of gender 'performativity' is based on the repetition of actions and practices which are sustained over time. Yet these performances are not spatially and temporally fixed (Evangelista, 2013) and can be deliberately modified (Butler, 1990). In this chapter, I reflect on my performance of Filipino masculinity in the field, and how this performed masculinity affects my interaction with research participants during my doctoral studies fieldwork.

I am a Filipino, in my late 20s and at the early stage of an academic career, currently working towards a doctoral degree at a New Zealand institution. I was preparing for fieldwork when one of this book's editors asked me if I would like to write a chapter for it. I thought: Why would she (Heike) think that I am capable of contributing a scholarly work on masculinities, especially when she knows that I am gay? Yet I came to realise that I oversimplify the attribution of the concept of masculinity with the male gender and the performance of male sexuality. Through my self-portrayal and behaviour, I realise that I portray a certain level of masculinity in my relationships with others and in my work. I narrate my experiences of negotiating Filipino masculinities, and my self-presentation and performance of masculine behaviour during field research in rural communities in the Philippines.

The Plot

Gender and gender roles are socially constructed. Masculinities, the ideologies attached to the characteristics of men, and their roles in society, vary according to social, economic, political, historical and cultural contexts (Rubio & Green, 2011). Before delivering a reflexive account of my gendered experiences during my fieldwork in the Philippines, it is important to provide a brief background of Filipino hegemonic masculinity and homosexual identities.

Like many colonised territories, centuries of Western colonialism have distorted traditional gender roles and ideologies in the Philippines. Pre-colonial societies in the archipelago had enjoyed 'sexually egalitarian' social structures (Alcantara, 1994). Women and men had equal participation in socioeconomic activities, and the demarcating line according to one's sexuality within such roles was not clearly identified (Eviota, 1994). Yet the introduction of Roman Catholicism through Spanish colonisation has embedded misogyny and patriarchy in native Philippine cultures (Dionisio, 1994), with consequent changes to gender roles and relations in the country.

Valledor-Lukey (2012) suggests that gender ideologies in present-day Philippine society are strongly attributed to the Spanish culture. This could possibly be because American colonisers were not successful enough in changing the social and political fabric of the Filipino way of life that they inherited from their predecessors, even though the Americans had put in place a more egalitarian colonial agenda than their Spanish counterparts (e.g. through providing a public school system that allows both girls and boys to attend). Although in terms of economic and workplace policies the Philippines has been revealed as one of the most 'gender equal' countries in the world today (McKinsey Global Institute, 2018), Filipino society is still largely considered patriarchal.

It has been observed that patriarchal ideologies are integrated in the roles expected of individuals in a Filipino family (Alcantara, 1994). Legally, the husband is prescribed as being responsible for supporting his family and as the 'administrator' of conjugal properties (Republic of the Philippines, 1949). Most commonly, the father is viewed as the *haligi ng tahanan* ('post of the house'). Traditional views on gender roles and identities imply that men should convey a strong character and image in society.

But what is Filipino masculinity? What are the characteristics that inform a Filipino man's *pagkalalaki* ('masculinity')?

Empirical studies, mainly with psychological approaches, have been conducted to address these questions in the context of contemporary Philippine society. Rubio and Green (2011) developed the Filipino Adherence to Masculinity Expectations (FAME) scale. Their findings revealed the following dimensions that underpin Filipino masculinity: 'assertiveness and dominance', 'family orientedness', 'sense of community', 'responsibility', 'integrity', 'intelligence and academic achievement' and 'respectful deference to women and the elderly'. Valledor-Lukey (2012) found more specific gendered characteristics of Filipino *pagkalalaki* and categorised them as either socially desirable or undesirable (Table 5.1). Taken together, these two scales suggest that the Filipino

Table 5.1 Filipino masculine traits

Positive traits	Negative traits
makapagkapwa ('affinity with others')	*mapusok* ('impetuous')
makisig ('elegant')	*matigas ang ulo* ('stubborn')
malakas ('strong')	*mayabang* ('proud, boastful')
maprinsipyo ('principled')	*padalus-dalos* ('hasty')
matapang ('brave')	
may kusang-loob ('with initiative')	

Source: Adapted from Valledor-Lukey (2012).

social constructions of masculinity revolve around the ideas of showing strength, proactive participation in community building, having strong values and principles and honouring women and the elderly.

Conversely, men who do not conform to hegemonic Filipino masculinity are susceptible to being labelled as gay. In other words, Filipino men who show effeminate behaviours and feminine traits, in terms of how they act, present themselves and treat/relate to others, may have 'suspicions' raised about their sexuality. For example, men who are caught physically hurting women are called out as *bakla* ('gay') and are usually confronted with the phrase, '*Kalalaking tao, nananakit ng babae. Bakla siguro?*' ('How could a man like him hit a woman. Maybe he is gay?'). In more traditional communities, men who find themselves doing jobs that are traditionally done by women may feel a sense of inferiority and tend to reference a degree of *kabaklaan* ('gayness') in performing such jobs. This was evident in a study of masculinity in a fishing village in the Philippines. Artisanal fisher folks who were forced to resort to 'fishmongering' due to declining fish stocks described the work as *nakakabakla* or 'work that turns one into a homosexual' (Turgo, 2014: 16).

Many Filipino scholars have deconstructed the notion of being *bakla* in the Philippines. Garcia (2004) suggests that the word *bakla* does not necessarily equate to homosexuality. As Diaz (2015: 721) explains, '*bakla* often denotes gay male identity, male-to-female transgender identity, effeminized or hyperbolic gay identity, and gay identity that belongs to the lower class'. A *bakla* is often regarded as a 'woman trapped in a man's body'. The idea of being *bakla* intersects with other gender identities because the *bakla* is also described as *pusong babae* or 'a man with a woman's heart' (Tan, 2001: 139). *Bakla* deviates from the ideal Filipino *pagkalalaki*, as observed in the derogatory use of the term illustrated in earlier examples. Nonetheless, being called out as *bakla* leads to social stigmatisation (e.g. Garcia, 2004; Tan, 2001).

Tolerance towards homosexuality in the Philippines was found to be one of the highest of the 'religious countries', yet the sexual marginalisation of the *bakla* in various social spheres is still evident in contemporary Filipino society. This 'minoritisation' of being homosexual can be traced from, and has been furthered by, the imposition of Roman Catholic and Christian values on Philippine society (Joaquin, 2014). The common connotation of being *bakla* is closely linked with the image of *parloristas* – gay men (usually effeminate, cross-dressers) who work in beauty salons – forming a hegemonic idea of Filipino male homosexuality especially within traditional rural communities, including the localities involved in my doctoral study. While various expressions of male homosexuality are increasingly tolerated in contemporary times and reproduced through the media (e.g. telenovelas, variety shows), wider acceptance of male homosexuality issues remains elusive (De Leon & Jintalan, 2018). This could be because such expressions are confined by cultural and societal views that prescribe family-friendly and patriarchal ideologies (Lee, 2002). It could also be that the masses are not fully aware of the multiple male homosexual identities (*mga uri ng kabaklaan*), which are often complex, multi-layered, contextual and, more importantly, individualistically defined.

The Stage and the Cast

My doctoral research explores the process and nature of community change induced by tourism and social entrepreneurship. The research design aims to give voices to the host communities who are regarded as the beneficiaries of the social enterprise projects involved in the study. My task was to gain insight into the outcomes of social entrepreneurship and tourism through the eyes of the community. The qualitative and community-centred nature of the research required me to be immersed in field study sites, to interview residents individually or in groups and to engage in informal interactions with the members of the community.

I report my field experiences and gendered performance in two rural communities in the town of San Felipe in the province of Zambales in the Philippines (Figure 5.1). These communities are small enclaves or settlements, within a *barangay* or village. Usually composed of a minimum of 40 families, they resemble small villages within a *barangay*. This means that social ties in these localities are usually based on kinship (i.e. *magkakamag-anak*) or friendship, indicating close social relationships between individuals within a small geographic boundary. My fieldwork in these two communities was performed from June to August 2018.

The first community heavily relied on fishing and farming as their main livelihood prior to becoming a popular surfing and tourism destination. Aside from its proximity to Manila, the community's popularity has been increased by the efforts of a social enterprise that has provided accommodation, surfing classes and other tours to visitors to the locality since 2011. My first impression of this locality was that it did not resemble a typical coastal community in the Philippines.

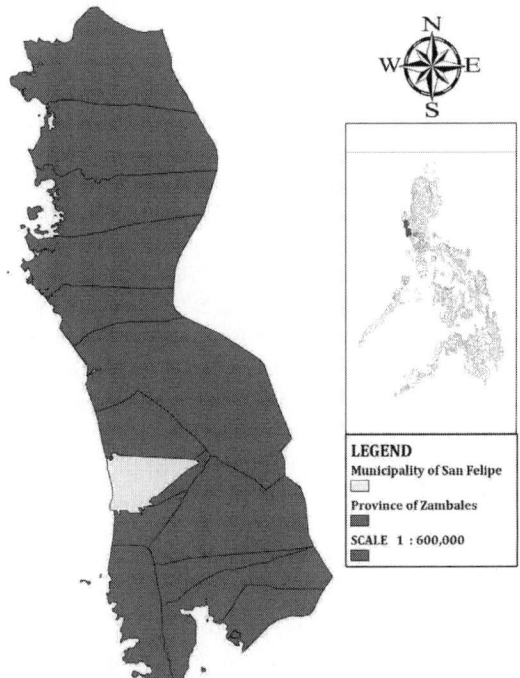

Figure 5.1 Location map of San Felipe, Zambales Province, Philippines
Source: Municipality of San Felipe (2013).

Although fishing boats and gear were stored in some residential yards, indicating that fishing was still practised by some community members, I observed that surfing culture had been largely assimilated into the local way of life.

The assimilation of surfing culture was evidenced by the number of surf shacks and surfing-based establishments that had mushroomed in the locality. Surfing had also become a popular leisure activity among locals, especially teenage boys. During the tourist season, the youth and younger men worked as surf instructors and took part in surfing competitions. Within the activities that bound this group of surfers, Filipino *machismo* was highly performed through *barkadahan* (or *barkada*), which is also known as *tropa* ('group of friends'). Angeles (2001: 13) likened *barkada* to a 'gang phenomenon' where Filipino males demonstrate 'homosociality and masculine solidarity' – hence a form of brotherhood. As in other surfing tourism locations, surfing-based socioeconomic activities in the first community were male dominated. In some aspects of local life, however, women also undertook important roles and positions in the community – the community's leader was a woman and some of the respected entrepreneurs in the community were women. However, patriarchal ideologies were still strong in the community, especially in family structures. For example, when women research participants were asked 'who is the head of the family?', all of them referred to their husbands, even in households where women appeared to undertake more paid employment than their male spouses.

The second community was engaged in a form of eco-cultural tourism in partnership with local government and a social enterprise. Since it was situated along mountain ranges, this community was more difficult to reach. It took about one and a half hours to travel from the town centre to the community, using a combination of three different transport modes: tricycle, *kariton* (a carabao-drawn carriage) and trekking.

At the time of the fieldwork, there were about 50 families living in the area. The community welcomed visitors during the weekends only, reserving weekdays for traditional livelihood activities. During the week, men went farming and hunting; women usually stayed at home to look after children and make products and souvenirs to sell to visitors during tour days (weekends). In terms of tourism-related roles, women were tasked with cooking food for visitors, taking care of a tree nursery and selling products and produce to tourists. Men were assigned to guiding and 'guarding' visitors, to ensure the latter's safety during visits. By looking at these roles, it can be construed that the community restricts its members to conventional gender identities, which are embedded in their local social dynamics. My initial observations and perceptions of gender ideologies in these rural communities shaped my later interactions and gender performances in the field.

Locating Myself in the Performance

To better situate myself in the performance, I had to reflect on my own identity and sense of masculinity. I had to continuously negotiate the performativity of my gender and sexuality in order to meet the standards of Filipino *pagkalalaki*. Against the backdrop of hegemonic *pagkalalaki* and societal attitudes towards gay male individuals in the Philippines, I will first articulate my journey and the experiences that shaped my sense of self and gay male masculinity.

I was born into a working-class family and raised in a rural *barangay* in a predominantly Catholic province in the northern Philippines. My father had to work overseas to support us, leaving the responsibility of raising me to my mother. Growing up, I heard family members and friends telling my mother to be 'cautious' about how she was raising me. They were worried that I might only acquire feminine characteristics growing up because my father was not always around. I was often told that my gestures were soft and effeminate compared to those of other boys my age. Naturally, my mother had to defend me and she told them that I was only acting soft and clumsy (*malamya*) because I was skinny compared to my peers.

Our *barangay* was traditional, with livelihood activities based on farming and small-scale aquaculture. Openly gay men (e.g. *parloristas*) and lesbian women (e.g. *tomboy*) in our community were often ridiculed and made fun of. My uncles threatened that if I they caught me making soft, effeminate movements and not acting masculine enough, they would place me in a sack and hang it up under a mango tree (until I finally learnt). Perhaps my family were just concerned that I might experience the treatment that effeminate *bakla* in the Philippines usually receive. In the Philippine context family reputation is important, especially 'in the eyes of the community' (Angeles, 2001: 10). I guess it is because of this environment and 'upbringing' that I learned to suppress my then feminine side.

Social pressures forced me to conform to the standards of Filipino *pagkalalaki*. I grew up being vigilant about not showing any effeminate movements or expressions, acting straight when I was in front of my family, friends and classmates. After I figured out my sexuality, I was in the closet (and still am today, at least when with family members). I did not want to disappoint my family, even though I was considered an 'achiever' and other people viewed them as 'lucky to have me'. It will be disappointing for my parents, and those others, if they find out that I am queer. Lee (2002: 40) explains that Filipino men are usually expected to form their own families and 'pressure becomes even greater as a homosexual tries to conceal his sexual identity from them [family members]'. Similarly, witnessing how the effeminate *bakla* is marginalised in everyday discourses and interactions, I wanted to grow up being respected by Filipino society by being competent at what I do while fulfilling traditional gender roles and images.

However, before graduating from university I thought that I should not hide any more, especially when entering the workplace. I also thought that in the tourism industry being gay is usually 'tolerated', but I was too idealistic. In my first job as a concierge officer in one of Manila's luxury hotels, I was tasked to work in a department that was mostly staffed by men aged mid-40s to 50s who exuded an aura of traditional Filipino *machismo*. My new colleagues were suspicious about my sexuality, and when they asked, I just told them the truth. During that time, I believed that expressing who I am was nothing but being fair to myself and the people that surround me. Yet their treatment of me gradually changed: inappropriate and indecent jokes around my sexuality began being thrown at me. I felt I was not being taken seriously at work. I left that job after over a year. I also promised myself that in my future workplaces my sexuality would not be others' business. Yet, to ensure this, I had to repackage my masculinity in a way that wouldn't show 'signs' of feminine behaviour (through movements, actions or language). Through being 'manly' enough, I intended to suppress suspicion and to be treated professionally at work.

This pledge to myself changed when I moved to New Zealand in 2013 to study for a master's degree. In that year, same-sex marriage had just been legalised in the country. I found myself in a space where my sexuality was not taboo and homosexuality is widely accepted. In my interactions with others, I acted in a way that expressed my identity and self-image – 'gay male that appears approximately masculine'. I have no hesitation about revealing who and what I am at work, at home and with my friends (most of them are gay Filipino men too). I feel that I have never been as comfortable and as safe in my gender identity as when I am in New Zealand.

My upbringing and social experiences in both the Philippines and New Zealand influence my negotiation of the different dilemmas surrounding my sexual orientation and identity. Living in a diverse Western society has made me abandon worries about whether I am 'manly enough' for other people, because this does not really matter. My sexuality (of *kabaklaan*) and the negotiation of its performance (of *pagkalalaki*) was best described by Diaz (2015: 721): '[sexuality is] conditional and contextual, and its deployment often points to the geographic, temporal, and material constraints'. These conditions apply whenever I return to the Philippines, especially when I perform professional and academic fieldwork: I have had to go back into the closet when doing field research in rural or, rather, 'traditional' Filipino communities.

Packaging My Masculine Self

Due to the negative experiences I have had in the Philippines concerning my gender and sexuality, my more relaxed and carefree self-expression when I am in New Zealand must be modified when returning to the Philippines for field research. My performativity adjustments need preparation. Like being a tourist (e.g. Hyde & Olesen, 2011), fieldwork is a performance that needs to be carefully packed for. My main priority was to prepare for the practicalities of conducting data collection and field research. Although packing for my gender performance is not explicitly linked to the latter, I know that, while in the field, I have to keep my masculinity in place in order to be respected and to establish my credibility as a researcher. Having this agenda in mind means packaging myself to approximately conform to the hegemonic image of Filipino *pagkalalaki* through my clothing, language, actions and interactions with others.

To be honest, I did not think about my gender performativity prior to immersing myself in the field. Also, like most researchers who are new to sites they are researching, I had limited knowledge about the study sites, except knowing that they may be traditional given the 'rurality' of their settings. In the Philippines, having a man's body means I have fewer rigid restrictions in terms of what attire to wear in public spaces, compared to women who may be at risk of being 'catcalled' if, for example, they wear less conservative clothing. I don't have a problem in terms of presenting myself through clothing, as my fashion sense leans towards masculine attire. However, like gender, clothing standards are also shaped by the stages we are acting on and the other actors we interact with, and looking 'professional' and 'appropriate' depends on the setting in which the social interactions occur. When meeting with public officials or visiting offices I sport smart casual attire, usually pants and a collared shirt. I do not wear the same outfit when I am in the communities, as potential participants may feel intimidated by a professional self-presentation. In rural communities, therefore, I normally wear a plain printed t-shirt matched with walk shorts, sandals and a cap on sunny days, or with a raincoat on rainy days (Figure 5.2). I think this is the most neutral way I can present myself to the community – not formal yet not too casual, and the attire most fitting to my study-site setting.

My major concern, however, is not 'looking' masculine but acting 'approximately masculine'. Perhaps staying at home for the first few days after arriving in the Philippines, before getting into the field, provided some time for me to 'acclimatise' and 'repackage' my masculine self. Language is a medium for self-expression, and for the Filipino *bakla* the use of gay lingo is a way of connecting with peers and expressing a 'gay'

Figure 5.2 The researcher

and 'jolly' self, which are usually desirable characteristics. Due to my covert homosexual identity, I am constrained from expressing myself in the way that I normally would. I have to disconnect from my true self when I am in the field.

The Performance

A number of steps must be undertaken in the process of my field research, namely: accessing the study sites; informal interactions in the community; approaching potential participants; and actual face-to-face interactions with study participants during individual and group interview sessions. Gaining access to the study sites required personal communication with the 'gatekeepers' of the localities such as local leaders and tourism officials. Usually, interactions with these individuals happened in professional settings (e.g. offices) bounded by formalities and protocols that I am familiar with. In such cases, I did not have any challenges in presenting my professional masculine image, as my training and work background had prepared me well for such engagements.

Arguably, being situated in the communities was the most challenging part of the fieldwork. For one thing, even though I am of Filipino descent, I could be considered an outsider (or *ibang tao*) in these localities. Establishing good social relations with local actors in communities is imperative in any field research endeavour. In the Philippine context, there are explicit social relationship stages that outsiders must enter into with community members, through informal interactions, in order for mutual trust to be generated between all those involved in a social activity such as research.

Establishing mutual trust and building rapport (*pakikipagpalagayang-loob*) was not easy (Aquino, 2019), especially if I was viewed as *ibang tao*. Reaching such a level of confidence meant I had to satisfy different modes/levels of social interactions – from basic civility (*pakikitungo*), to mixing (*pakikisalamuha*), to more complex forms of conforming (*pakikibagay*) and adjusting (*pakikisama*; Enriquez, 1979, 1986; Santiago & Enriquez, 1976) – through informal engagement with the communities. I spent most of my time in the field, specifically staying at a local hostel in the surfing destination community. In my first weekend on site, I mainly assumed the role of a visitor. The need to experience what it is like to be a tourist in the area required me to participate in social and tourist activities. Drinking culture is quite prominent in the locality. In the hostel that I stayed at, drinking was seen as a social activity and the medium through which friends are gained. I had to partake in this activity even though I am not into this kind of drinking, because I planned to stay in the hostel and needed to show my host that I could mix with other people and that I was not a KJ or 'kill joy'. However, participating in mixing and mingling activities was not limited to within the fences of the hostel as I also had to engage both formally and informally with the locals as part of my field research.

One of my goals was to appear as masculine as I could through my behaviour and actions. I observed that on lean tourism days, drinking is a popular leisure and socialisation activity among the locals. One day, I found myself asking for information from a group of drinking *barkada*; I was not able to leave their vicinity until I had consumed a shot of gin with *calamansi* juice as the chaser. I had to *tagay* or 'shot' even though I was not in the mood to drink. In my view, undertaking such an action was a way to avoid being labelled as *walang pakikisama* (someone who cannot adjust to the local way of life) or, worse, *bakla*. On another night, I was walking down a narrow alley when a pack of dogs barked at me and approached me in a threatening way in front of locals who then tamed the dogs. During the incident I stayed calm and walked straight on even though I was screaming and dying inside. I had to act in such a way in order to conform to 'masculine' behaviour, illustrate that I have affinity towards others and show bravery, all of which are known as characteristics of Filipino men (Valledor-Lukey, 2012).

Patriarchal and masculine ideologies were prominent in the rural communities too. My local guide's son fell from a carriage while we were on our way to another research location. As the boy was about to cry, the guide comforted his son with the words:

Huwag kang iiyak, lalaki ka eh. Dapat hindi umiiyak. ('Don't cry, because you're a man. You should not be crying.')

I also observed that being gay was stigmatised in the community. One afternoon I was walking on the beach and watching skim boarders. I heard playmates teasing a boy who had become their laughing stock because he was accused of 'hanging out with a *bakla*'. These instances made me conscious of how I was behaving during those times. Was I masculine enough? I was reminded of when I was not open about my sexuality, particularly of how anxious I was about monitoring my own actions and whether others perceived my gender performance as masculine enough (e.g. Chan, 2016).

Despite these field anxieties, there were also times when I felt successful in concealing my gay male identity while in the field. As was evident during interview conversations, some male study participants were at ease in making 'boys' comments and jokes with me. A married male interviewee commented when I told him that his wife could also join the interview session:

Tayo na lang dalawa mag-usap, huwag na siya isama. Baka may malaman pa siya. ('Let's just talk, the two of us. Don't let her join because she might know something [she is not supposed to].')

One of the interviewees who worked as a surfing instructor narrated that surfing was his and his friends' way to attract 'chicks'; as he told me:

Dito, hindi mo na kailangang umalis para humanap ng chicks. Ikaw na pupuntahan dito. ('Here, you do not have to go out and look for girls. Here, they come to you.')

Personally, I prefer not to engage in these kinds of conversations because I am not used to having them within my social circles. However, I realised that these individuals were at ease in talking about these topics with me, because they viewed me as someone who could relate to their performed masculinities.

Conclusion

This chapter is probably the most difficult piece that I have written thus far, because the narratives presented here are personal. As an early-career researcher and an academic-in-training, my usual focus is to get the job done. In the past, I have not really reflected on how my gender performance and expression are situated in doing research.

As already explained, I am aware that gender roles are socially and culturally constructed. Drawing on my personal experience, I have had to negotiate my own gendered identities depending on where I was physically located (e.g. the Philippines or New Zealand) or who I was interacting with (e.g. family, friends, workmates). In my most recent fieldwork experience in the Philippines, I had to recalibrate my masculine identity when engaging with individuals in the field. My reason was simple: to be treated with basic civility, respect and professionalism. I had to present my masculine self due to the traditional background of the communities involved in my research, because homosexuality and various *kabaklaan* may not be well tolerated in such settings.

I believe that presenting myself in such a way was beneficial to my purpose. I did not encounter a negative experience, at least with my study participants. I have a good level of confidence about the data I have collected from individuals because I think they were comfortable interacting with me. To be honest, what concerned me more was the psychological distress I occasionally experienced. Performance anxiety hit me when I encountered sexist masculine jokes and stigmatising homosexual comments, or when I felt that I was not conforming well to the local masculine image (e.g. Rubio & Green, 2009).

Others who read this account may feel that I was not being true to myself in terms of expressing who I am. Yet, as in any other job, we must wear certain masks when we are in the field, in order to establish rapport and form professional relationships with our research participants (without deceiving them), although I also think that it would have been easier to connect to my participants, especially straight men, if I was a straight male individual (I guess I will never really know). As I have stated earlier, my gender and sexuality are my business. I have agency in deciding which masculine identity(ies) to express; it is likewise for anyone else finding themselves in the same dilemma. However, in my case I may have to continue maintaining a masculine image in the field until homosexuality is more accepted in the Philippines where, so far, most of my research projects have been situated.

I hope this chapter has provided some background on Filipino masculinities, and contributed to the limited knowledge on the influence of Asian masculinities in doing and being in field research (e.g. Rawat & Khoo-Lattimore, 2016). By articulating how I negotiated my male homosexual identities in the field, this chapter has also responded to the call for Asian and early-career academics to embody reflexivity especially in conducting qualitative research (Khoo-Lattimore, 2018; Mura & Khoo-Lattimore, 2018). I hope that this chapter will provide insights for other researchers who find themselves in the same personal, social and cultural circumstances.

References

Alcantara, A.N. (1994) Gender role, fertility and the status of married Filipino men and women. *Philippine Sociological Review* 42 (1), 94–109.

Angeles, L. (2001) The Filipino male as 'macho-machunurin': Bringing men and masculinities in gender and development in the Philippines. *Kasarinlan Journal of Third World Issues* 16 (1), 9–30.

Aquino, R.S. (2019) Towards decolonising tourism and hospitality research in the Philippines. *Tourism Management Perspectives* 31, 72–84. doi:10.1016/j.tmp.2019.03.014

Butler, J. (1990) *Gender Trouble: Feminism and the Subversion of Identity*. New York: Routledge.

Chan, J. (2016) 'Am I masculine enough?': Queer Filipino college men and masculinity. *Journal of Student Affairs Research and Practice* 54 (1), 82–94. doi:10.1080/19496591.2016.1206021

De Leon, J.A. and Jintalan, J. (2018) Accepted or not: Homosexuality, media, and the culture of silence in the Philippine society. *Jurnal Komunikasi, Malaysian Journal of Communication* 34 (3), 408–425. doi:10.17576/jkmjc-2018-3403-25

Diaz, R. (2015) The limits of *bakla* and *gay*: Feminist readings of *My Husband's Lover*, Vice Ganda, and Charice Pempengco. *Signs: Journal of Women in Culture and Society* 40 (3), 721–745.

Dionisio, E.R. (1994) Sex and gender. In E.U. Eviota (ed.) *Sex and Gender in Philippine Society: A Discussion of Issues on the Relations between Women and Men* (pp. 1–34). Manila: National Commission on the Role of Filipino Women.

Enriquez, V.G. (1979) Towards cross-cultural knowledge through cross-indigenous methods and perspective. *Philippine Journal of Psychology* 12 (1), 9–15.

Enriquez, V.G. (1986) Kapwa: A core concept in Filipino social psychology. In V.G. Enriquez (ed.) *Philippine World-View* (pp. 6–19). Singapore: Institute of Southeast Asian Studies.

Evangelista, J.A. (2013) On queer and capital: Borrowing key Marxist concepts to enrich queer theorizing. *Philippine Sociological Review* 61, 349–370.

Eviota, E.U. (1994) The social construction of sexuality. In E.U. Eviota (ed.) *Sex and Gender in Philippine Society: A Discussion of Issues on the Relations between*

Women and Men (pp. 53–82). Manila: National Commission on the Role of Filipino Women.

Garcia, J.N.C. (2004) Male homosexuality in the Philippines: A short history. *IIAS Newsletter* 35, 13.

Hyde, K.F. and Olesen, K. (2011) Packing for touristic performances. *Annals of Tourism Research* 38 (3), 900–919. doi:10.1016/j.annals.2011.01.002

Jagger, G. (2008) *Judith Butler: Sexual Politics, Social Change and the Power of the Performative.* London: Routledge.

Joaquin, A. (2014) Carrying the cross: Being gay, Catholic, and Filipino. *Sociology and Anthropology Student Union Undergraduate Journal* 1, 17–25.

Khoo-Lattimore, C. (2018) The ethics of excellence in tourism research: A reflexive analysis and implications for early career researchers. *Tourism Analysis* 23 (2), 239–248. doi:10.3727/108354218X15210313504580

Lee, R.B. (2002) Psychosocial contexts of the homosexuality of Filipino men in heterosexual unions. *Journal of Homosexuality* 42 (4), 35–63. doi:10.1300/J082v42n04_03

McKinsey Global Institute (2018) *The Power of Parity: Advancing Women's Equality in Asia-Pacific.* See https://www.mckinsey.com/featured-insights/gender-equality/the-power-of-parity-advancing-womens-equality-in-asia-pacific

Municipality of San Felipe (2013) *Comprehensive Land Use Plan.* San Felipe: Office of the Municipal Planning and Development Coordinator.

Mura, P. and Khoo-Lattimore, C. (2018) Locating Asian research and selves in qualitative tourism research. In P. Mura and C. Khoo-Lattimore (eds) *Asian Qualitative Research in Tourism: Ontologies, Epistemologies, Methodologies, and Methods* (pp. 1–20). Singapore: Springer Singapore.

Rawat, K. and Khoo-Lattimore, C. (2016) The impact of masculinities in the researcher–respondent relationship: A socio-historical perspective. In C. Khoo-Lattimore and P. Mura (eds) *Asian Genders in Tourism* (pp. 53–64). Bristol: Channel View Publications.

Republic of the Philippines (1949) *Republic Act No. 386: The Civil Code of the Philippines.* See https://www.wipo.int/edocs/lexdocs/laws/en/ph/ph021en.pdf

Rubio, R.J. and Green, R.-J. (2009) Filipino masculinity and psychological distress: A preliminary comparison between gay and heterosexual men. *Sexuality Research and Social Policy: Journal of NSRC* 6 (3), 61–75.

Rubio, R.J. and Green, R.-J. (2011) Filipino men's roles and their correlates: Development of the Filipino adherence to masculinity expectations scale. *Culture, Society and Masculinities* 3 (2), 77–102. doi:10.3149/csm.0302.77

Santiago, C. and Enriquez, V.G. (1976) Tungo sa maka-Pilipinong pananaliksik. *Sikolohiyang Pilipino: Mga Piling Papel* 1 (4), 3–10.

Tan, M.L. (2001) Survival through pluralism: Emerging gay communities in the Philippines. *Journal of Homosexuality* 40 (3), 117–142.

Turgo, N.N. (2014) Redefining and experiencing masculinity in a Filipino fishing community. *Philippine Sociological Review* 32, 7–38.

Valledor-Lukey, V. (2012) Pagkababae at pagkalalake (femininity and masculinity): Developing a Filipino gender trait inventory and predicting self-esteem and sexism. Unpublished doctoral dissertation, Syracuse University.

6 How Masculinity Creeps In: Awkward Field Encounters of a Male Researcher

Can-Seng Ooi

Ethnographic research does not follow a clear path of theoretical conceptualisation, data collection, data analysis and report writing. The process engages all these research components simultaneously. The process is ongoing. While there are debates on what ethnography is (Brewer, 2000; Gobo, 2011; Hammersley & Atkinson, 2007), this is not the place to review the discussions. Broadly, I take an inclusive understanding of ethnography, including the focus on rich case studies and the collection of data through various qualitative methods that involve deep engagement with participants and the community such as observations, participant observations and in-depth interviews. Documents, public information and other relevant sources of information also matter in bringing about a more holistic understanding of the community and the social phenomena studied. Anthropology is the underlying discipline, and ethnographic analysis is holistic in nature and requires the field researcher to critically self-reflect on their own encounters and experiences. Writing this chapter forces me to reflect deeper into my masculinity in research. My various masculine expressions are embedded in the entirety of the ethnographic research process, and that process reveals how I have negotiated the field with participants and my own self-perceptions.

This chapter is divided into three sections. After the introduction, I will explain why I am using an awkward encounters methodology to reveal how I have managed, suppressed and flaunted my masculinity in the field. Masculinity here means what it means to be a man (Vanderbeck, 2005: 387). Being a man is, however, a social process. My field experiences affirm the understanding that there are a variety of masculinities, and the expression, assertion and constitution of masculinity is situated and negotiated (Vanderbeck, 2005). By reflecting on several awkward encounters, I uncover how I have claimed and redefined my various situated masculine fieldworker identities. The following section then focuses on three field encounters, discussed in the context of the active interviewing method, of developing a holistic understanding of the field and of self-reflexivity in my analyses. The discussion and concluding section takes on an awkward question posed by Crick (1995) on the similarities between anthropological fieldworkers and tourists, raising questions about not only the validity of ethnographic research but also whether it is possible for fieldworkers to fully understand the studied culture and society. I asked myself: Am I just like a male tourist when I do my fieldwork?

Awkward Encounters in the Field

As mentioned in the opening paragraph, the ethnographic process is not a linear one. It is also one that accentuates contexts and circumstances, while embracing a holistic approach to understanding society and culture by the researcher. The data collection process involves various methods, including observation, participation and interviews. These qualitative methods acknowledge the presence and influence of the fieldworker. My masculinity creeps into how I collect my data. And during and after my data collection I reflect on the data and situate them in context; this reflexive analytical process allows my masculinity to creep in too. The spinning wheel in Figure 6.1 accentuates the interrelationship between data collection and analyses in ethnographic work. The researcher is central in that process.

I have an embodied problem. My masculinity is visible even though it is not usually discussed in the field or reflected in my writing. My masculinity is taken for granted. Making the tacit explicit is an important part of ethnographic research because the tacit matters, such as how nuances are communicated by greetings and the subtle switch in linguistic codes. Masculinity and being a man have been reflected in many studies (e.g. Jarvis, 2015; Liong, 2015; Treadwell, 2015; Vanderbeck, 2005; Vorobjovas-Pinta & Robards, 2017) but more often than not, masculinity can be intentionally and tacitly ignored and marginalised. My male identity intersects with other aspects of me. And it was an encounter that made me

Figure 6.1 How masculinity creeps into my research: the spinning wheel of ethnographic research

realise that I was inadvertently (or otherwise) trying to suppress my gender and sexuality in showcasing my fieldworker identity! This was what happened.

I dress appropriately for fieldwork, following social norms. For instance, I dress more formally when I talk to government officials and more casually to meet participants who have become closer. In one instance, I met a gay activist in Singapore to seek his perspectives on gay tolerance in the city-state and how the promoted tolerance has been used to support tourism and the city's creative economy strategy (Ooi, 2018). I did not inform him about my sexuality as I did not want to make my homosexuality the basis of why I was doing the project. Tan (all participant names are not real in this chapter) and I met at a café. To my horror, we were wearing the same clothes – tight grey T-shirts and faded blue jeans. It was awkward. We had coffee and held our discussion. The discussion did not go as smoothly as I hoped. In hindsight, I should have joked about our common dress sense at the start, and that might have alleviated my sense of discomfort. But why was I uncomfortable?

There were many reasons, but I realise that the main one was the 'threat' to my professional image as an objective researcher. Just by association with Tan – in this case our meeting and our coincidental dress sense – I felt that I was inadvertently outed. Misguided or otherwise over the years, I have deliberately tried not to reveal my sexuality to participants until they are friends or when they ask. Reflecting further, I have also intentionally ignored my embodied masculinity; I do not want to be seen as a male fieldworker, just as a professional fieldworker. The awkward encounter with Tan forced me to reflect. I was hoping to portray myself as an asexual, gender-neutral researcher. In the name of professionalism and being objective, my sex/gender and sexuality should not be seen to affect my work. Neither should my race/ethnicity. Writing this chapter allows me to or rather forces me to confront my masculinity and the intersections of my fieldworker identity.

Ethnographic descriptions and analyses are necessarily selective. Unless it is the focus of the research, masculine expressions are left tacit and not focused upon. To do otherwise may result in the danger of over-dramatising or overanalysing 'irrelevant' issues. But masculinity in the field is not an irrelevant issue, especially with the prevalence of male privilege and the perpetuation of 'hegemonic masculinity' in social structures, institutions and norms (Beasley, 2008; Buscatto, 2011; Cheng, 2008; Connell & Messerschmidt, 2005; Donaldson, 1993). My attempt at being a gender-neutral, asexual professional fieldworker is an example of how I have internalised a view that my gender is neutral. In contrast, as I will show later, I feel a need to claim my professional authority by compensating for possible negative perceptions because of my race/ethnicity.

To aid me in this chapter, I revisited notes from some of my awkward field encounters, following the methodology advocated by Koning and

Ooi (2013). Awkward encounters in fieldwork are common. Despite the celebration of ethnographic work as being contextual and being explicit about what fieldworkers do in the field, ethnographies are also written with the aim of depicting authority and objectivity. The presentation of emotions that seem to question the motive of the fieldworker or their ability to control themselves are normally ignored, so that the soundness of the research will not be questioned (Burkitt, 2012). Emotions of anger, fear, embarrassment, helplessness and ambivalence often cast doubts on our ability to think straight and to be objective (Kleinman & Copp, 1993: viii; van Maanen, 2010). Instead, the fieldworker is presented as the hero. Inconveniences and challenges in the field are presented in thick descriptions but they are eventually handled and managed. Unresolved embarrassing, frustrating and helpless moments are often ignored in the writing. This is unfortunate because these emotional and awkward encounters will reveal and help us understand underlying social structures that we have internalised. Following criticisms of reflexivity as too rational (Burkitt, 2012) and as a 'means' to claim better research (Pillow, 2003), confronting awkwardness offers an 'inclusive reflexivity' (Koning & Ooi, 2013). Inclusiveness here means the admission of issues that fieldworkers often block out because their studies may appear less credible and reliable.

This chapter is written with the goal of taking a few of my awkward field encounters seriously and revealing the masculinity embedded in my fieldwork. This process is itself uncomfortable for me but these encounters provide 'analytical clues' (Whiteman, 2010: 334). They are also 'epistemologically informative' (Davies, 2010: 13) encounters that make a 'productive difference' (Alvesson *et al.*, 2008: 495) in generating a richer appreciation of the human experience (Cunliffe, 2003).

Framed another way, there are at least three reasons why awkward encounters matter in ethnography and fieldwork. Firstly, doing ethnographic fieldwork is an active process. The fieldworker influences and affects the outcomes. This has to be acknowledged even when the outcomes are not positive. The field is a dynamic environment; the ethnographical presentation reflects reality more honestly when discomforts are also documented.

Secondly, ethnography recognises the joint presence of the researcher and the participants in the field, and more. A holistic understanding of the situation requires not only the incorporation of complementing, contrasting and contradictory interests and agendas of various parties into the analysis but also situating the unfolding social actions in the structural and institutional context. Awkward encounters reveal structural and social institutional underpinnings that invoke those emotions.

Thirdly, critical self-reflexivity reveals our own politics in fieldwork, writing and our own sense of self. We are not just being selective in our writing but also in what we observe and how we feel in the field. Reflecting on our mixed feelings make us clearer about our own true positions. For this chapter, the process will also be contrasted with the circumstances that I feel comfortable in, leading to further reflections which will expose the tacit advantages and privileges I get for being a male fieldworker.

Masculinity Creeps into the Field

In this section I will present three awkward encounters within the context of doing ethnographic work. Each encounter is tied to three central tenets of ethnography, namely the active and responsive interviewing method, developing a holistic understanding and critical self-reflexivity.

Active and responsive interviewing: Masculinity and intersected identities

Any interview situation is social: the interviewer and respondent interact. I have always used and acknowledge the 'active interviewing method' (Holstein & Gubrium, 1997). In that interaction, participants incite the production of meanings that address issues relating to the questions I ask. This is different from the perspective that the interview conversation is framed as a potential source of bias, error, misunderstanding or misdirection, i.e. a persistent set of problems to be controlled. The corrective is then to get the interviewer to ask questions in a fixed manner, so that the respondent will give out the desired information (Holstein & Gubrium, 1997: 113–120). But that ignores the biases of responses to the identity of the interviewer: gender, seniority, appearance, etc. Even in my conscientious attempt at acknowledging the dynamic dialogue between the participants and me, I was obsessed (as discussed earlier) by how I tried to marginalise my masculinity and sexuality in the name of being a professional fieldworker. My fieldworker identity intersects with other identities.

Being a man matters, but there are other aspects of my being that 'rank higher' than my masculinity. In one of my most uncomfortable situations in fieldwork, I brought two friends/students to the artist village at Songzhuang, China. They were interested in art and joined me for my field visit. The visit was part of my ongoing project on art worlds, tourism and the creative economy. A more detailed account of that field visit is presented in Koning and Ooi (2013). Regardless, my two friends – a white Danish woman and a Middle-Eastern looking French man – generated a set of social interactions that I felt uncomfortable with because I was not in control of the field situation. I had been to Songzhuang many times, visited different studios and spoken to various artists. This time, Jing befriended us while we were on the public bus into the town. Jing's English was halting and I interpreted between him and my friends, Adly and Susan. As the day wore on, it was clear that he thought that I was their guide, even though I had told him that Adly and Susan were my MBA students from Denmark. These dynamics provided me with a set of observations that I could not have done without Adly and Susan's presence. Despite my polite attempts at correcting the inaccuracies, Jing was using our presence to flog his self-proclaimed overseas connections and credentials to his friends and colleagues around the artist village as he showed Adly, Susan and me around.

I could not convince Jing that I was Adly and Susan's professor, or possibly he decided to ignore my correction. My perceived role as a guide was also assumed when I went shopping with my Scandinavian executive students (mostly in their 30s and 40s) at the various tourist-friendly malls in Shanghai and Beijing. Whenever I was with them I did not get much service and attention, as they – male and female – were perceived as wealthy tourists; I was always seen as their guide and interpreter. It is clear that my racial appearance and ethnicity define how many people in China see me when I am with my European friends and students.

In fieldwork, I am also hoping to draw out appropriate responses from participants. The active interviewing method sees the interview as neither an innocent process nor a source of distortion: it is a 'site of, and occasion for, producing reportable knowledge itself' (Holstein & Gubrium, 1997: 114). In the presentation of self, I want to be seen as a professional researcher that transcends gender, sexuality and ethnicity. Nonetheless I realise that my appearance as East Asian affects my encounters in the field. I am cognisant of how I have sometimes been relegated to the role of a guide rather than a researcher, or of my credentials being negatively affected because of my ethnicity. So I always try to contextualise my professional status by mentioning where I work and my position at the university. As I have become older and more senior in the academic system I have become more confident in approaching participants, assuming that looking older and being a professor is an easier 'sell'. Also, mentioning Copenhagen and Copenhagen Business School when I was living and working there seemed to have drawn positive responses. Now that I am living in Tasmania and working at the University of Tasmania, I also mention that. Inadvertently at the first encounter, I am responding to how I imagine others perceive me. Before a closer relationship of trust develops, I want to exert more authority. While riding on a hegemonic masculine regime, I compensate for my ethnic identity 'shortcomings'. My fieldworker identity is not only situated, it is one that is negotiated within myself by how I see that others imagine me.

In pursuit of holism from a masculine perspective

Holism is a central tenet in ethnography. It is a principle that affects how we collect data and analyse data. The issue of holism, when pushed to the limit, can also paralyse social analysis (Marcus, 1988). And it creates the fear of not having done enough fieldwork or analysis. All social phenomena entail an extensive understanding in ethnography. Such a project would also defeat research if one could not decide which issues and processes are more relevant and significant. While I have been balancing selectivity and holism in my research, writing this chapter makes me wonder if I have a 'masculine holism', that is, a form of holistic understanding dictated by a masculine slant.

In one of my trips to M50, a popular tourist spot and arts district in Shanghai, I brought along a colleague, Anne. I wanted to introduce her to the area and share with her my art and tourism research on China. She is Scandinavian, in her 20s then, blonde and attractive. That visit was very different from my countless other visits to M50. Gallery owners and artists were less interested in talking to me; I was treated as her guide and interpreter. One artist and gallery owner was particularly drawn to her. I had visited his studio many times before and we had held conversations, but this time I was ignored. He was more interested in talking to Anne, using me as the interpreter. He said that he liked to draw beautiful girls, and instantly drew a silhouette of Anne. Anne was uncomfortable. I felt that I had to protect Anne and tried to leave the space. We did that soon after Anne received the drawing. Anne and I spoke about the experience. We speculated on how her views of China and the Chinese were shaping with the unwanted attention. For me, that particular experience made me feel 'left out', 'used' by the artist and 'redundant' in the social interaction.

I have not managed to integrate that experience (and other similar ones) into my analyses. Neither have I stopped inviting fellow researchers and friends into the field with me. In the context of this chapter, as the art world is layered with many gatekeepers and informal social relations, I wonder how my masculinity has tainted my views. I am tempted to say that I have maintained a disinterested, objective and unbiased holistic view of tourism and the art world, but could I have a different view if I were a blonde lady? I have written about how the art world helps launder money and how friendship matters in becoming a successful artist, but I have not been able to speak about (hetero)sexual tensions in the space of doing and selling art (Ooi, 2010, 2011, 2017). But sex matters, as already analysed critically by female scholars (see accounts in Porter & Schänzel, 2018). My masculine holism understanding is surely limited, taking on a largely male 'objective' perspective.

Masculine reflexivity: Lessons from shades of awkwardness

Ethnography is selectively presented. There are implications for excluding from our texts the 'awkward' moments we have encountered in our research. Koning and Ooi (2013) argue that reflexivity has become too rational-cognitive and fails to account for our emotions. As such we engage with the 'embodied practice' of reflexivity (the doing); however, this is unmistakably intertwined with the 'textual practices' (Alvesson et al., 2008). Reflexivity is connected to the crisis of representation and legitimation, and that challenge was already present in Malinowski's time. Foley (2002) argues that Malinowski's fieldwork diaries come close to what we would refer to as confessional reflexivity. Today's reflexivity – reflecting on the research process, the interaction between researchers and researched, the interpretation of data and the subsequent knowledge claims – supposedly incorporates what confessional reflexivity did not yet do. It supposedly encompasses the research community as a whole, including the site and others in the network such as peer reviewers (Hardy et al., 2001). And reflecting on awkward encounters is yet another step towards adding another layer to producing better ethnography. But in which direction should our reflexivity go?

Similarly and yet in contrast to my 'left out' experience with Anne, I have another very awkward field situation which shows that I do not know how to frame and direct my reflexive thoughts. It is an incident that still haunts me. I was accosted in the field once, by a woman (Jane). It occurred when I was in a bookshop in Singapore, as I was exploring the art books as part of my tourism and art world research. A lady asked me to reach up for a book on the top shelf. I did so and we struck up a conversation. She asked me if I was an artist. Jane said she painted herself. I took the opportunity to tell her about my interest in understanding Singapore's art world and tourism, and that I meet artists to find out how they navigate Singapore's art world. I asked her about her practice and, when the opportunity came about, I asked if I could interview her. She eagerly agreed and asked if I wanted to visit her studio immediately. I could not but suggested the next day. Jane then gave me her calling card. I went to her studio. While there she intimated to me that her husband was wealthy. She then further elaborated that her husband was not treating her well and had been cheating on her. She revealed very personal information about herself. After about 30 minutes, I realised that she was propositioning me. My immediate reaction was to sound very formal, while I (insensitively?) continued to ask her questions and find out about her art practice. I did not feel threatened, just uncomfortable. Jane was eventually exasperated, and asked me directly, 'What's wrong with me? Are you married?'. The situation was awkward. I made my exit politely, and did not follow up with her.

It was a new situation to me, and my spontaneous defence was to focus on my research rather than on my safety or the appropriateness of the unfolding events. I did not feel threatened. Many if not most female fieldworkers have had to negotiate advances from men (Godfrey & Wearing, 2018; Hamilton & Fielding, 2018; Martinez & Peters, 2018; Munar, 2018). But for me, such a fear is quite alien. That has to be male privilege.

My ethnographic reflections have a strong masculine tinge, if not coloured accordingly. Reflexive methods have undermined positivist perspectives by revealing that interpretations and negotiations are needed to re-contextualise observation situations at all junctures of fieldwork and analysis (Baszanger & Dodier, 1997; Clifford & Marcus, 1986; Hopper, 1995). Since one's background and the working context cannot be avoided, then all scientific analysis are eventually loaded interpretations. And for me, my seeming fearlessness (or naïveté) stems from my physical ability and also my masculine worldview. I was insensitive. Being empathetic is essential in reflexive work, as we ponder on the circumstances and are self-critical. This incident exposed to me that I was not empathetic towards a sexually charged situation so as to respond in a quicker (and more morally appropriate) manner.

Discussion and Conclusion: A Personal Response to a Professional Crisis

The discussion above has focused on three basic tenets of ethnography: the responsive interviewing process; developing holistic analyses; and being critically self-reflexive. They all reveal how I have taken my masculinity and male privileges for granted. While I negotiate my fieldworker identity, I am also trying to assert authority, legitimacy and objectivity in my self-presentation. Being male helps. There is a tendency for me to compensate for perceived shortcomings in my fieldworker identity such as ethnicity and age, while tacitly exploiting the privileges that come with being a man.

Putting on analytical lenses to highlight my tacit masculinity has been an eye-opening process. It is also an awkward process, as I confront myself and wonder if I have the authority and ability to discuss hegemonic masculinity and male privilege. This exercise reminds me of an argument by Crick (1995). Crick poses an awkward question to tourism researchers: In what ways is the anthropologist studying tourism like or unlike the tourists being studied? (Crick, 1995: 205).

Crick argues that anthropological fieldworkers and tourists are distant relatives. The role of these researchers and tourists overlap in at least two aspects. Firstly, to residents, fieldworkers are likely to be classified and treated like tourists. Secondly, tourist interests and those of a researcher are increasingly similar. Tourists and researchers acquire local knowledge and analyse the place. Processes of culture, social change, meanings and symbols of the local community have become the interests of tourists, just like researchers. These two observations run parallel to seeing the male tourism fieldworker as similar to a male tourist.

There is an inevitable gap between tourists and their understanding of the host community, despite their efforts and claims about wanting to engage with residents: (1) tourists do not have local knowledge; (2) their visits are short; and (3) their intention is to enjoy themselves (Ooi, 2002: 18–20). The male fieldworker is like a male tourist as he acquires local knowledge, visits for a limited period and intends to seek information that may lead to publications. Furthermore, the male fieldworker, like the male tourist, comes with his masculine worldview and engages with the host society from the perspective of his own background, agenda and interests (see articles in Thurnell-Read & Casey, 2015). As male tourists look at the destination through their own lenses, would not the male researcher do the same? For instance, male tourists take on riskier adventures and activities and are even expected to do so (Lozanski, 2015; Thurnell-Read, 2015). As a male fieldworker, I have few personal security fears in my field encounters.

Writing this chapter has created a professional and existential problem of sorts for myself. No research project is perfect. There is a reason why we try to address the politics of academic knowledge. In spite of my self-criticism of trying to be professional and authoritative, and consequently trying to marginalise my gender, sexuality, ethnicity and the like, I am still situated in a hegemonic masculinity regime. Our embodied background matters but how should that be addressed? Can I ever be a good researcher?

We should avoid an essentialist response to the fact that all research is inevitably gendered. Gender matters and I am certain that fieldworkers are cognisant of this and are capable of addressing known biases. There are at least three arguments to differentiate the researcher from the tourist, and to address gender discrimination.

The first is that fieldworkers are not sightseers but are more detailed and serious in their encounters, regardless of their embodied gender. They are obliged to be systematic, holistic, historical and contextual in understanding the situation. Errington and Gewertz (1989: 46) argue that ethnographers differentiate their intentions and the politics of their visits, in contrast to those of tourists. Tourists often do not have adequate understanding of social issues and the world political economy, and many are not interested in these (Errington & Gewertz, 1989: 51). Although many tourists are no longer ignorant of the world they travel in nor uninterested in learning more, researchers seek relevant and wider academic knowledge. Academic analysis, interpretations and studies on social structures and institutions are more rigorous than anecdotal and cursory tourist experiences. From a gender perspective, hegemonic masculinity as embedded in social structures and institutions should be more directly and explicitly addressed. It can be argued that gender biases will always exist or are difficult to change, but researchers should not surrender and agree to perpetuate them in analyses. Tourist perceptions, understanding and analyses of the host do not have the rigorous grounding necessary to be distributed as scientific knowledge (and are accordingly discounted).

Another argument to differentiate tourists from researchers is that researchers have 'deeper' encounters with the society than tourists with the destination. Male and female fieldworkers have had countless genuine encounters with the other sex over the course of their lives, and have understood their family and friends as individual human beings. These experiences – although embedded in hegemonic masculinity – would inform their fieldwork. It remains that the fieldworker can never be totally integrated into the community, and a male fieldworker cannot comprehend the embodied experiences of a female. To claim otherwise is like saying that tourists can become native. Regardless, self-reflexivity and the ability to admit the limits of the fieldwork experience are helpful in building up trust in the analysis; the fieldworker is honest and does not oversell their research. Conscience and responsible tourism have taken hold (Jamal, 2019), but fieldworkers are trained to be more critically self-reflexive of the situation. In this context, the male fieldworker should put on gender lenses to interrogate the biases and injustices they may enjoy (or suffer). As I mentioned earlier, it was not usual for me to explicitly reflect on gender biases in my analyses, but the preparation of this chapter has led me to think otherwise and to layer my interpretations with a set of interrogations on my own masculine experiences.

The final argument to differentiate researchers from tourists is through a moral claim. Some researchers see their activities as more worthy than tourists' activities. Tourists and economic tourism, on the other hand, can harm the destination. Local culture is touristified and commodified (Cohen, 1988; Munar, 2018; Ooi, 2005). Tourists are said to be indulging in hedonistic neo-colonialism, collecting souvenirs, photographs and other things they find of symbolic value to them, which they transform into status back home. Fieldworkers want to help the community, and conscientiously try not to affect or destroy it. Crick reminds us that researchers are also engaged in 'scientific colonialism' (Crick, 1995: 210). Researchers collect data for their research, so that they become publishable products to advance their careers (Crick, 1995: 210–212). While no fieldworker would intentionally do 'evil', we do perpetuate hegemonic masculinity. This moral differentiation between tourists and researchers can only hold water if we also address gender issues.

At the start of this chapter, I recount my attempt at being professional and objective. My encounter with Tan revealed my identity politics: I was ignoring my gender and sexuality to portray myself as a professional researcher. And with my reflections on other awkward encounters, have I found a way to detach myself from my masculinity and become a neutral fieldworker? No. What I could only do is to: (1) explicitly acknowledge my own masculine presence as an element in the research process; (2) reflect on the research process and interrogate the male privilege I enjoy; and (3) consciously acknowledge gender biases and address masculine hegemony in my structural and institutional analyses. My holistic view of the world will always be tainted, and I can only try to go beyond masculine holism.

Quality is important in scientific work. Awkward encounters are marginalised in ethnographic studies because we want to appear objective and scientific. But each ethnographic presentation has to be evaluated on its own terms. Institutionalising the presentation of awkward moments would increase the scientific worth of ethnography because uncomfortable situations do not only exist, they matter in how fieldworkers affect the situation. Research would benefit if social ambiguities and emotional ambivalence are evaluated and integrated into our research.

References

Alvesson, M., Hardy, C. and Harley, B. (2008) Reflecting on reflexivity: Reflexive textual practices in organization and management theory. *Journal of Management Studies* 45 (3), 480–501. doi:10.1111/j.1467-6486.2007.00765.x

Baszanger, I. and Dodier, N. (1997) Ethnography: Relating the part to the whole. In D. Silverman (ed.) *Qualitative Research: Theory, Method and Practice* (pp. 8–23). London: Sage.

Beasley, C. (2008) Rethinking hegemonic masculinity in a globalizing world. *Men and Masculinities* 11 (1), 86–103. doi:10.1177/1097184X08315102

Brewer, J. (2000) *Ethnography*. Buckingham: Open University Press.

Burkitt, I. (2012) Emotional reflexivity: Feeling, emotion and imagination in reflexive dialogues. *Sociology* 46 (3), 458–472. doi:10.1177/0038038511422587

Buscatto, M. (2011) Using ethnography to study gender. In D. Silverman (ed.) *Qualitative Research: Theory, Method and Practice* (3rd edn) (pp. 35–52). Singapore: Sage.

Cheng, C. (2008) Marginalized masculinities and hegemonic masculinity: An introduction. *Journal of Men's Studies* 7 (3), 295–315. doi:10.3149/jms.0703.295

Clifford, J. and Marcus, G.E. (eds) (1986) *Writing Culture: The Poetics and Politics of Ethnography*. Berkeley, CA: University of California Press.

Cohen, E. (1988) Authenticity and commoditisation in tourism. *Annals of Tourism Research* 15 (3), 371–386.

Connell, R.W. and Messerschmidt, J.W. (2005) Hegemonic masculinity: Rethinking the concept. *Gender & Society* 19 (6), 829–859. doi:10.1177/0891243205278639

Crick, M. (1995) The anthropologist as tourist: An identity in question. In M.-F. Lanfant, J.B. Allcock and E.M. Bruner (eds) *International Tourism, Internationalization and the Challenge to Identity* (pp. 205–223). London: Sage.

Cunliffe, A.L. (2003) Reflexive inquiry in organizational research: Questions and possibilities. *Human Relations* 56 (8), 983–1003. doi:10.1177/00187267030568004

Davies, J. (2010) Introduction: Emotions in the field. In J. Davies and D. Spencer (eds) *Emotions in the Field: The Psychology and Anthropology of Fieldwork Experience* (pp. 1–31). Palo Alto, CA: Stanford University Press.

Donaldson, M. (1993) What is hegemonic masculinity? *Theory and Society* 22 (5), 643–657. doi:10.1007/BF00993540

Errington, F. and Gewertz, D. (1989) Tourism and anthropology in a post-modern world. *Oceania* 60, 37–54.

Foley, D.E. (2002) Critical ethnography: The reflexive turn. *International Journal of Qualitative Studies in Education* 15 (4), 469–490. doi:10.1080/09518390210145534

Gobo, G. (2011) Ethnography. In D. Silverman (ed.) *Qualitative Research: Theory, Method and Practice* (3rd edn) (pp. 15–34). Singapore: Sage.

Godfrey, J. and Wearing, S. (2018) Negotiating machismo as a female researcher and volunteer tourist in Cusco, Peru. In B.A. Porter and H.A. Schänzel (eds) *Femininities in the Field: Tourism and Transdisciplinary Research* (pp. 23–36). Bristol: Channel View Publications.

Hamilton, J. and Fielding, R. (2018) Safety first: The biases of gender and precaution in fieldwork. In B.A. Porter and H.A. Schänzel (eds) *Femininities in the Field: Tourism and Transdisciplinary Research* (pp. 10–22). Bristol: Channel View Publications.

Hammersley, M. and Atkinson, P. (2007) *Ethnography: Principles in Practice* (3rd edn). London: Routledge.

Hardy, C., Phillips, N. and Clegg, S. (2001) Reflexivity in organization and management theory: A study of the production of the research 'subject'. *Human Relations* 54 (5), 531–560. doi:10.1177/0018726701545001

Holstein, J.A. and Gubrium, J.F. (1997) Active interviewing. In D. Silverman (ed.) *Qualitative Research: Theory, Method and Practice* (pp. 113–129). London: Sage.

Hopper, S. (1995) Reflexivity in academic culture. In B. Adam and S. Allan (eds) *Theorising Culture: An Interdisciplinary Critique after Postmodernism* (pp. 58–69). London: University College London Press.

Jamal, T. (2019) *Justice and Ethics in Tourism*. London: Routledge.

Jarvis, N. (2015) Masculinity and the Gay Games: A consideration of hegemonic and queer debates. In T. Thurnell-Read and M. Casey (eds) *Men, Masculinities, Travel and Tourism* (pp. 58–72). London: Palgrave Macmillan. doi:10.1057/9781137341464_5

Kleinman, S. and Copp, M. (1993) *Emotions and Fieldwork*. London: Sage.

Koning, J. and Ooi, C. (2013) Awkward encounters and ethnography. *Qualitative Research in Organizations and Management: An International Journal* 8 (1), 16–32. doi:10.1108/17465641311327496

Liong, M. (2015) Like father, like son: Negotiation of masculinity in the ethnographic context in Hong Kong. *Gender, Place & Culture* 22 (7), 937–953. doi:10.1080/0966369X.2014.917280

Lozanski, K. (2015) Heroes and villains: Travel, risk and masculinity. In T. Thurnell-Read and M. Casey (eds) *Men, Masculinities, Travel and Tourism* (pp. 28–42). London: Palgrave Macmillan. doi:10.1057/9781137341464_3

Marcus, G.E. (1988) Contemporary problems of ethnography in the modern world system. In J. Clifford and G.E. Marcus (eds) *Writing Culture: The Poetics and Politics of Ethnography* (pp. 165–193). Berkeley, CA: University of California Press.

Martinez, E. and Peters, C. (2018) Gender bias and marine mammal tourism research. In B.A. Porter and H.A. Schänzel (eds) *Femininities in the Field: Tourism and Transdisciplinary Research* (pp. 109–125). Bristol: Channel View Publications.

Munar, A.M. (2018) Researching in a men's paradise: The emotional negotiations of drunken tourism fieldwork. In B.A. Porter and H.A. Schänzel (eds) *Femininities in the Field: Tourism and Transdisciplinary Research* (pp. 170–184). Bristol: Channel View Publications.

Ooi, C.-S. (2002) *Cultural Tourism and Tourism Cultures: The Business of Mediating Experiences in Copenhagen and Singapore*. Copenhagen: Copenhagen Business School Press.

Ooi, C.-S. (2005) State-civil society relations and tourism: Singaporeanizing tourists, touristifying Singapore. *Sojourn – Journal of Social Issues in Southeast Asia* 20 (2), 249–272.

Ooi, C.-S. (2010) Cacophony of voices and emotions: Dialogic of buying and selling art. *Culture Unbound* 2, 347–364.

Ooi, C.-S. (2011) Subjugated in the creative industries: The fine arts in Singapore. *Culture Unbound* 3, 119–137. See http://www.cultureunbound.ep.liu.se/v3/a11/cu11v3a11.pdf.

Ooi, C.-S. (2017) The global art city. In J. Hannigan and G. Richards (eds) *The SAGE Handbook of New Urban Studies* (pp. 207–216). Thousand Oaks, CA: Sage.

Ooi, C.-S. (2018) Global city for the arts: Weaving tourism into cultural policy. In T. Chong (ed.) *The State and the Arts in Singapore* (pp. 165–179). Singapore: World Scientific. doi:10.1142/9789813236899_0008

Pillow, W. (2003) Confession, catharsis, or cure? Rethinking the uses of reflexivity as methodological power in qualitative research. *International Journal of Qualitative Studies in Education* 16 (2), 175–196. doi:10.1080/0951839032000060635

Porter, B.A. and Schänzel, H.A. (eds) (2018) *Femininities in the Field: Tourism and Transdisciplinary Research*. Bristol: Channel View Publications. doi: https://doi.org/10.21832/PORTER6508

Thurnell-Read, T. (2015) 'Just blokes doing blokes' stuff': Risk, gender and the collective performance of masculinity during the Eastern European stag tour weekend. In T. Thurnell-Read and M. Casey (eds) *Men, Masculinities, Travel and Tourism* (pp. 43–57). London: Palgrave Macmillan. doi:10.1057/9781137341464_4

Thurnell-Read, T. and Casey, M. (eds) (2015) *Men, Masculinities, Travel and Tourism*. London: Palgrave Macmillan. doi:10.1057/9781137341464

Treadwell, J. (2015) The lads just playing away: An ethnography with England's hooligan fringe during the 2006 World Cup. In T. Thurnell-Read and M. Casey (eds) *Men, Masculinities, Travel and Tourism* (pp. 189–203). London: Palgrave Macmillan. doi:10.1057/9781137341464_13

Vanderbeck, R.M. (2005) Masculinities and fieldwork: Widening the discussion. *Gender, Place & Culture* 12 (4), 387–402. doi:10.1080/09663690500356537

van Maanen, J. (2010) A song for my supper. *Organizational Research Methods* 13 (2), 240–255. doi:10.1177/1094428109343968

Vorobjovas-Pinta, O. and Robards, B. (2017) The shared oasis: An insider ethnographic account of a gay resort. *Tourist Studies* 17 (4), 369–387. doi:10.1177/1468797616687561

Whiteman, G. (2010) Management studies that break your heart. *Journal of Management Inquiry* 19 (4), 328–337. doi:10.1177/1056492610370282

Part 3
Situated Masculinities

7 A Tale of Two Researchers: Masculinity in Cross-cultural Contexts

Joseph M. Cheer and Alan A. Lew

> Those humans socialised as men must be more meaningfully exposed to the benefits of gracefully 'descending' while more women 'ascend' – the idea that underpins the book's title. Men, yield not only your political and material forms of power and privilege (seats in parliaments, corporate leadership positions, domestic decision-making), but find yourself some gentle, non-defensive ways to retreat less anxiously from the punishing edicts of legacy masculinity, and you may find contentment.
>
> Kagan, 2017: 29, on The Descent of Man
by Grayson Perry, 2016

Introduction

Contemporary discourses of masculinity are framed by dramatic shifts in the context of emergent and mostly more audible feminist discourses that agitate for a necessary reshaping of gender relations, as alluded to in the quotation above. This is underlined by the bottom-up #metoo social justice movement, on the one hand, and the seemingly anachronistic attitude of Australian Prime Minister Scott Morrison in his 2019 International Women's Day address where he emphasised that the rise of women should not come 'at the expense of men', on the other. Debates over how femininities and masculinities might best coexist, or not, resonate amid so-called 'gender wars' and the pursuit of a more equal and productive steady-state, as captured by Kagan (2017) above. The term 'gender wars' itself alludes to tensions and struggles that are more than theoretical and very much steeped in everyday life, defining and delineating how female, male and other exchanges and encounters are moderated and constructed. Indeed, the question of gender, and the archetypal attributes and constraints inherent and assigned accordingly, have evolved into multiple categorisations that increasingly question simplistic notions, moving discussions into unprecedented territory both for scholarly social sciences and within mundane everyday contexts.

Engaging in the full suite of gender-related theoretical and empirical discourses is beyond the scope of this chapter. However, acknowledgement of this as an all-encompassing theme is vital given that both masculinities and femininities in scholarly practice draw from and link back to the diverse and voluminous works contained in gender theory and related feminisms and their application. Judith Butler's (1990) *Gender Trouble: Feminism and the Subversion of Identity* was seminal and instructive in underlining contemporary entreaties concerning gender and identity and the related performativity of gender and the way this moulds attitudes towards gender relations. Her work foreshadowed understandings that any elicitation of gender recognises that binaries and norms must give way to more complex and elastic conceptualisations. As Butler (1990: vii) quipped: 'I sought to counter those views that made presumptions about the limits and propriety of gender and restricted the meaning of gender to received notions of masculinity and femininity.'

The line of questioning underlining the reflections offered here foregrounds how our performance of masculinity in field research has been negotiated and shaped by cross-cultural contexts. Cross-cultural contexts very often present a rethinking of how masculinities (and ethnicities) are performed and negotiated, especially where the research site is in great contrast to that normally inhabited by the researcher. In the same way, patriarchal research contexts tied to conceptions of masculinity have underlined our work, and these often privilege male researchers while at the same time placing demands and expectations on how they ought to perform and negotiate their masculinity. Indeed, when the gendered context of a research site is at odds with or alien to our acculturated norms and values, how do we as male researchers reconcile personal, ethical and moral dilemmas that may arise? The rising call for reflexivity on such issues is emblematic of fieldwork encounters that many male scholars experience when working both within and outside their home base. However, it is especially pronounced in cross-cultural settings where differences can be very stark. When in the field, such encounters become everyday situations that are too easily taken for granted. For social scientists like us, our ability to engage, shape and (re)construct our research endeavours is profoundly influenced by these gendered performances. How we practise and play out our masculinities requires an ongoing and intentional process of awareness, reflexivity and adjustment.

Accordingly, we recognise Butler's contributions to gender discourses and ruminations and acknowledge her conceptualisations to underline how we conceive of masculinities – not in objection to femininities, but rather as complementary and abiding notions in theory and more so in practice, and apropos of the performance of gender in the everyday. This accords with British artist Grayson Perry's *uber*-modern conceptualisations of masculinity as tending towards embodying greater diversity and complexity: 'masculine values, such as stoic self-sufficiency, can be chronically damaging to men, and can harm their relationships with their partners and with themselves' (Haig, 2016). Indeed, Perry's contribution to thinking around masculinity, although mainstream in its phrasing, suggests that the evolution of masculinity is evidently a long, slow process:

> The attributes men are encouraged to aspire to – power, authority, rationality, competition, success, and no sissy stuff – are far more destructive than they are productive or reparative, and hark back to eras bygone (when they were no less toxic). (Kagan, 2017: 27)

In a sense, contemporary conceptualisations of masculinity are heavily problematised and draw from transformations of what Bennet and Sani (2004) refer to as the social self – that is, masculinity is socially constructed, and deliberations about a new and pervasive iteration are brought about by transformations that are largely personal and local. The meanings and symbolisms inherent in the performativity attached to masculinity today reflect shifts in the social self, which in turn has an abiding influence over the extent to which masculinities and femininities converge or depart, and whether they overcome the binaries that pit them against each other – often as problematic bedfellows. Indeed, we equally acknowledge Clatterbaugh's (1998: 25) assertions that 'while some think that masculinities are biologically grounded, it is generally agreed that they are socially and historically constructed'.

Method

The principal and underlining aim mobilised here draws from autoethnography conducted by two researchers: the first researcher is at a middle career stage and the second at a more advanced, senior stage. The employment of autoethnography implies that researchers draw from a long period of engagement in the field, in which reflexivity is used to underline conceptualisations of broader social and cultural phenomena (Chang, 2016). Questions about research rigour and reliability are often levelled at proponents of autoethnography; however, in arguing for its legitimacy and dependability, Chang (2016: 10) points out that 'autoethnography is affirmed as an ethnographic research method that focuses on cultural analysis and interpretation'. In tourism research, the employment of autoethnography is largely muted, although becoming increasingly more common (Botterill, 2003; Everingham, 2016; Noy, 2008; Scarles, 2010).

Research Context

In articulating the fieldwork for this chapter of two authors whose body of work spans contrasting time periods, duration and geographies, the predication is that comparative perspectives can illuminate the extent to which the performance of masculinities in academic fieldwork contexts can differ, and how inherent challenges might be negotiated. For example, how might the first author's experience starting out in the 2000s contrast or converge with the second author's entrance into academic research in the 1980s? Moreover, to what extent have temporal, structural and conceptual constraints in the performative aspects of masculinities endured and how have these shaped respective researcher trajectories?

When it comes to the research context, we also dwell on *how the transformation of self occurred* for authors and *what underlying personal, familial, cultural and environmental influences have the potential to shape research practice and the performance of masculinity*. We also engage the line of questioning that Patterson and Elliott (2002) deploy: *How did we negotiate our own characteristics of male identity and masculinity in our research practice?* Additionally, we grapple with Mutua's (2013: 363) assertions that *in invoking masculinities in research, we are compelled to develop frameworks that 'more fully capture the complexities of men's lives'*.

Both researchers embody diverse backgrounds, based in different countries and starting out on their respective research practice at different times. However, commonalities lie in both having multi-ethnic identities (Asian and European plus Polynesian for the first author), while grounding their respective research practice in Western, developed country contexts (Australia and the United States, respectively). Ultimately, however, we hone in on fieldwork situations outside our respective home bases, especially in the Asia Pacific region, where cultural differences and normative understandings are evident and tend to depart from those we embody and identify with in our home context. *In particular, we outline negotiation strategies that serve to optimise fieldwork practice and participant engagement, alongside containment of fieldworker poise, sensitivities and reactions.*

In leveraging our research practice, which for both researchers is steeped in the Asia Pacific, we concur with Cornwall and Lindisfarne's (2016: 12) proposition that '[t]he many different images and behaviours contained in the notion of masculinity are not always coherent: they may be competing, contradictory and mutually undermining'. We maintain that this is central to much fieldwork, especially in cross-cultural contexts where patriarchal structures underline extant socialities.

For male researchers, cross-cultural contexts that challenge normative understandings of masculinity and maleness, and especially the appropriate placement of women in society, can be confronting, discomforting and alien. This is particularly the case when our notions of masculinity are a departure from that encountered in the field. Negotiating dissonance between fieldworker attitudes towards masculinities and the superstructures around what it means to be a man, alongside contrasting notions evident in cross-cultural fieldwork encounters, sows the seeds for a range of emotions including bewilderment, discontent, apprehension and abhorrence, among others. This tends to occur where *in situ* masculinities are severely at odds with what researchers are usually accustomed to.

For male researchers in the field, cross-cultural and entrenched patriarchal contexts can be challenging, especially where the performance of masculinities encounters symbols and behaviours that they would ordinarily consider close to or precisely unacceptable. Maintaining objectivity and aplomb amid confronting and conflicting situations challenges fieldworker practice and necessitates the development of considerations and contingencies that help negotiate for optimised rather than compromised researcher-participant interactions and subsequent research outcomes. Thus we also acknowledge Vanderbeck's (2005: 387) assertion of the need to explore 'the negotiation of masculinities during fieldwork, with an emphasis on issues confronting male researchers who fail to conform to dominant expectations of "manliness" which have currency in a given setting'.

Researcher 1

Background

> Being a real man isn't something 'internal,' but something performed – for other men. Masculinity is 'homosocial' – meaning it's other men who judge whether we're doing it right. We want to be a 'man's man,' not a 'ladies' man.' A 'man among men.' (Kimmel, n.d.)

Kimmel's assertions are doubtless designed to rouse but are clearly widespread sentiments when it comes to assessing what being a man is supposed to encompass. That masculinity is homosocial is rather representative of what Kimmel argues are anachronistic ideas about masculinity, yet still are pervasive in the present-day social milieu. While Kimmel's references are generalised, how masculinities are enacted by researchers in the field is inevitably linked back to the sentiments imbued in the opening quotation to this chapter. By extension, the question of what it means to be a male researcher is elicited and the extent to which masculinities in professional contexts are tied to generalised and well-codified ideas about masculinity and being a man in domestic and private settings are also put forward. Indeed, what qualities should underline male researchers in the field for optimum exchanges and what shapes understandings of masculinity in professional practice as opposed to the domestic, mundane everyday are further pathways for examination.

If my performance of masculinities in the everyday, and especially in my daily research practice, is shaped by social and historical contexts, it is fair to say that my particular circumstances and upbringing have had a fundamental influence on me as a person, but most importantly in my professional practice as a researcher. If we firstly consider the personal: in spending my formative years outside Australia where I live, growing up in Fiji where cultural, linguistic, religious and ethnic pluralism are predominant, cross-cultural encounters were part of the everyday (Hermann & Kempf, 2005). Small island developing states, especially those in the Pacific, are usually, although not altogether, dominated by patriarchal or matriarchal frameworks. On balance, Fiji maintains a stronger patriarchy by virtue of the roles that men and women play in some of the more vital institutions in government and in business. That men have a greater tendency to be Chiefs, or CEOs or military leaders, was apparent in my childhood and the status quo more or less remains the same (Chattier, 2015; Leckie, 2005).

At a familial level, my paternal make-up has its genesis in the flood of migrants that fled southern China in the 1800s, spreading out across the Asia Pacific region (Ali, 2005). My maternal inheritance is composed of what is colloquially referred to as half-caste, or *afikasi*, in which a mix of Pacific Islander and European extraction is evident (Mann, 1939). In my case, a stronger link to Rotuma, an island to the north of the Fijian archipelago and composed of Polynesians, was formative for me (more so than the Danish heritage of my maternal grandfather).

Furthermore, within my childhood family context, domestic discussions were based around the controversies generated by my father, a 'full blooded Chinese man', entering into marriage and starting a family with my mixed-race mother. On both sides of the family there was much consternation. On the Chinese side, my father had broken with family convention by eschewing an arranged marriage with a bride from mainland China. On my mother's side of the family, her cohabiting with a 'wily Chinaman' raised all sorts of worries and objections, especially given the double jeopardy of his being a Catholic as well, as the brand of religion followed on my maternal side was Jehovah's Witness. The result was a long-term estrangement from the paternal Chinese side and a continually simmering dissatisfaction on the maternal side underlined by criticism and suspicion.

Further to the potted family history above, it is evident that the upshot of a mixed-race and multicultural background has had a profound influence on my construction of self and the attitudes and perspectives attached to it. Linguistically, the use of English, Cantonese, Rotuman, Fijian and Hindi in my upbringing was normal, further reinforcing the ordinariness of multicultural frameworks in the everyday. On reflection, it is clear that what was normal were polysemic frameworks that neither conflicted nor confused but instead provided a lens through which my childhood self viewed the world. This is underlined by a sense that self and identity are not singular or static but instead emphasised pluralism and chameleonic performance in the everyday.

The subsequent family move to Australia in my early youth could not have been more different, with an evident chasm between what my childhood was comprised of, towards a monocultural and largely European/British context where multilingualism was uncommon. Evidently, this led to overwhelming questions about my identity and self and where in this melange my principal identity lay. With the passage of time, it has become clear to me that identity and self are less distinguished by definitive bifurcations that pigeonhole, and instead are more aligned to a multifariousness that was in step with the contemporary multicultural context in the Asia Pacific and urban Australian setting. Combined, these underlying influences have helped shape my personal outlook, and equipped me with a plasticity that underpins my adaptability and nimbleness in the practice of research, and shapes my performance of masculinity across the various sites where I ply my academic trade.

Researcher 1: In the field

To address the aims of this contribution, I will illustrate my research journey and focus on the formative period when my doctoral research was carried out. It was by accident rather than design that my doctoral fieldwork took me back to the Pacific islands where I was raised. Prior to making the leap across to academic life, I spent a period of time working with the Australian government as part of its international development efforts in Vanuatu (see Figure 7.1), specifically to boost the development of sustainable tourism. It was therefore a rather logical decision to continue this work, although not as a practitioner but instead as an academic conducting ethnographic fieldwork on the impacts of tourism.

Figure 7.1 Author in the field at Pentecost Island, Vanuatu
Source: Cheer (2011).

Having spent a considerable time living and working in Vanuatu, I had secured the social accoutrements that would enable me to seamlessly conduct fieldwork. I understood the sociocultural milieu intimately and had become fluent in the lingua franca of the country, *Bislama* – a creolised version of English spoken widely. Most importantly, I was quickly embraced by many of the participants of my research on account of my Pacific heritage, privileged by a bestowed Pacific brotherhood and defined in opposition to the fact that I was not a *'waet man'* – Bislama for 'white man of European descent' (the term European is used throughout the Pacific to describe white Americans, Australians, New Zealanders and French nationals, among others). My linguistic dexterity and comfortableness within the urban, rural and outer island setting gave me a head start when it came to kicking off my ethnographic fieldwork.

In Vanuatu, the performance of masculinity is largely representative of masculinities in Melanesian contexts more generally, where an abiding dominance of the patriarchy in indigenous institutions as well as in faith networks and in the military and judiciary persists (Taylor, 2008). However, I acknowledge Taylor's warnings about oversimplifying masculinities in Melanesia:

> ... any study of changing Pacific 'masculinities' in non-Western contexts should rightly be recognised as an instance of theoretical imperialism. (Taylor, 2008: 127)

One of Taylor's key references is anthropologist Marshal Sahlins, whose (1963) work on Melanesian socialities saw the construction of masculinities as being dominated by 'big men' – prominent and well-placed men who were adept at leveraging their positions for personal gain. During the course of my fieldwork, the gender position I encountered privileged male authority, where women, although valued for their contributions to public and private life, dominate the keeping of the hearth and as domestic managers and primary caregivers. This meant that, although I sought a more representative sample of community attitudes, ingrained socialities subordinated potential female participants, meaning that the vast majority of participants in my fieldwork were men; this made the space I inhabited an overwhelmingly male domain. This was also at odds with my examination of the impacts of tourism and the changes acting on traditional cultural frameworks in the country, because the greater majority of participants in the sector are women.

Illustrative of this is that early on during my doctoral fieldwork, I was to conduct a community meeting in the village of Anelgauhat on the island of Aneityum. I had become accustomed to only meeting men and my time in Aneityum was no exception. On arrival at the airport at around midday on the adjacent Mystery Island, the chairman of the local tourism association came to meet me. His main assistant was also a man and when I arrived at my bungalow in Anelgauhat, another man welcomed me even though it seemed clear that the woman in the background, his wife, was largely responsible for the smooth running of the bungalow. In fact, the entire tourism association was comprised of males, with the chairman apparently gaining his authority via patriarchal and Chiefly lineage.

On the afternoon of my arrival, I was introduced to the 'important' people on the island, such as the tourism association executive and other key people in the handful of enterprises in the village including the bank outpost, storekeeper, teachers and medical infirmary staff. All the people I had met were men with the exception of the community nurse. Later, I was taken to the *nakamal* for kava (see Figure 7.2). The *nakamal* is a traditional meeting place and a male domain where kava, a traditional social lubricant, is consumed and where local-level politics is discussed (Huffer & Molisa, 1999). However, the *nakamal* context is a largely male domain and if this is indeed where local politics is discussed, this disadvantages women in the community who find themselves at the fringes of decision making.

Prior to the commencement of the community consultation meeting, villagers gathered on the grassed area overlooking the foreshore at Anelgauhat Village. What struck me at once amid the random seating pattern that developed was the voluntary placement of women at the fringes of the gathering (Figure 7.3). Given my long association living and working in the country, this was unsurprising. However, as a researcher this presented a challenging set of circumstances and raised two main practical questions. How do I ensure that the voices of women in the community are heard and promote an inclusive consultation? How do I as a man behave so as not to undermine the confidence of male members of the community who had invited me to address the entire community?

Figure 7.2 *Nakamal*, Port Vila, Vanuatu
Source: Cheer (2011).

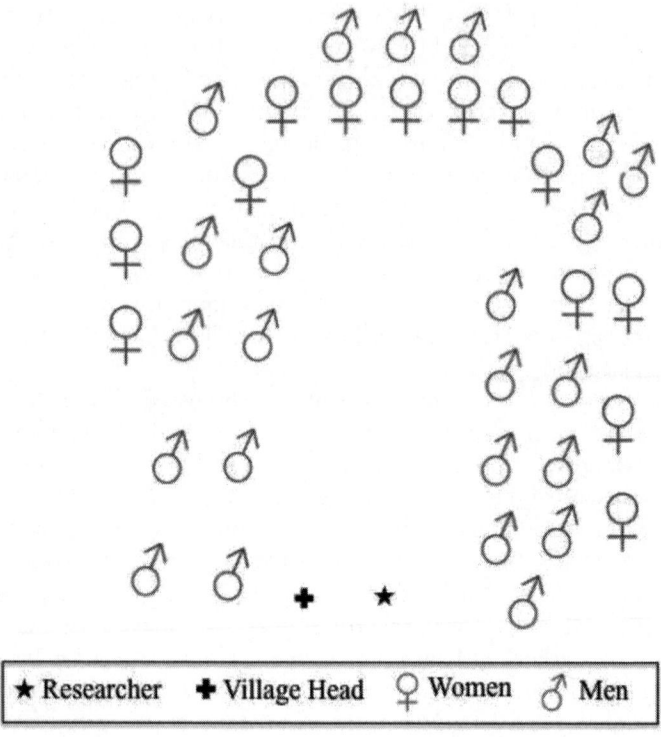

Figure 7.3 Community consultation meeting seating arrangement
Source: Author.

This scenario underlined the circumstances that shaped the way the practice of my masculinity would play out. Such situations where the researcher's interpretations of masculinity are at odds with those of participants are not unique but are magnified when the researcher is an outsider with little understanding of the sociocultural nuances and customary roles played by either gender. However, given my familiarity with Vanuatu and most of the cultural norms, I was attuned to the need for flexibility and was open to adapting my research practice. As a self-defined social progressive, my immediate instincts are to upend the dominance of men in the community and to seek out a more proportionate interaction with women and those outside traditional hierarchies. Amplifying the voices of women is vital in the pursuit of sustainable development and sustainable and inclusive tourism, and this aim was central to my modus operandi during the community consultation that unfolded.

How then did I secure the amplification of women's voices during the community consultation? How did this impact the practice of my masculinity amid a research context that had clearly divergent attitudes and norms? Throwing open questions to the gathering would have resulted in men largely responding with women remaining mute throughout. This was not going to be representative of community attitudes and would therefore compromise the rigorous application of this research. Furthermore, in attempting to correct the disproportionately limited voice of women at such gatherings, what right did I as an outsider have to 'rectify' such an imbalance? Most importantly, what would my key informants, who were all male, make of my actions? Would they consider this disrespectful to them or just an oddity?

In what was to follow, a series of decisions I made centred on how I conceived of my own maleness and masculinity, doubtless shaped by my particular social and cultural constructs. These are practical constraints that all researchers face in situations where the *in situ* culture is very different from what they are accustomed to. However, when male researchers are expected to conform to the local (and patriarchal) context, challenging these norms can be risky. Herein was the dilemma I faced in the demonstration of my masculinity: Should I go along with the structures in place or try to bend them towards what I saw was most acceptable? The instinctive decision was to bow to the patriarchal frameworks because that was the most practical thing to do. I felt that what would be most costly would be losing the good faith that I had developed with the key informants. I also felt that rectifying gender empowerment concerns through this one-off situation would be futile in the face of overarching frameworks and that I probably had more to lose as far as my research was concerned.

Consequently, the community consultation meeting proceeded as any other such gathering would and with most, if not all, of the discussion dominated by male attendees. For the most part, the women sat at the fringes listening attentively but deferring to their male counterparts when questions were raised. Male attendees also conferred among themselves during the meeting with little direct involvement of their female counterparts. It was only at the end of the meeting, when approaching female attendees directly, that I received some level of response. However, these were largely superficial and not in direct response to issues discussed.

In this vignette of my fieldwork and my associated conceptions of what I assumed masculinity to be (both mine and that *in situ*), I downplayed my own socially constructed normative understandings to abide by what was customary at the field study site. However, I initially encountered mild moral anxiety, given that as a socially progressive male I am highly cognisant of the femininities–masculinities binaries that exist and see the merits in addressing the impacts of disproportional patriarchal dominance. In a sense, this was a practical fieldwork response because I had made research effectiveness my priority rather than ideological leanings. The question as to whether my research practice was at odds with my own personal attitudes lingered, and this was reflected extensively in my field notes. For example, the following question to myself tended to resonate throughout my field notes: *Why are the voices of women less amplified than that of the men? Who speaks for the women? How do I encourage women to speak up without compromising them and my wider research?*

As an example of this, on many occasions during my doctoral fieldwork, more often than not in group situations, female participants would defer to their male counterparts to have first right of reply to my questions. This was often an unspoken norm and, where articulated, the response from female participants was: 'we should let the men talk first'. Ordinarily, pursuing representativeness in research assumes that I am able to canvass views that are reflective of the sample population, yet hearing male voices predominantly didn't sit too well with my personal views and, most importantly, with my professional research practice. Negotiating this delicate occurrence on a daily basis predominated my fieldwork practice and required tact insofar as how I projected and employed my masculinity when in the company of groups comprised of both men and women.

'What else I could have done?' was a line of questioning that swirled around my head for some time afterwards. As individual researchers, is it within our personal and professional remit to rectify overarching societal frameworks that we find disagreement with? In this case, it was the extent to which I was responsible for ensuring the female voice at the research site was amplified and not muted, as is more often the case. Was I contributing to the prolongation of a situation that downplayed the potential contributions of women? Or, in choosing not to push back against the established orthodoxy, was I selling out? Undoubtedly, moral and ethical dilemmas are not uncommon in research practice, yet little in my fieldwork planning and training had prepared me for this.

Researcher 1: Implications

On reflection, this fieldwork vignette has multiple implications for researchers sharing similar personal traits and fieldwork contexts, which include being male, raised in an affluent developed country, trained in a Western academic context, and used to institutions and thinking that promote affirmative action for women and girls set against the dominance of patriarchal frameworks. In Australia where I reside, gender issues and the empowerment of women are at the forefront of social change and still have some way to go to be rectified. This is in contrast to a country like Vanuatu where traditional patriarchal constructs are still very much influential over extant socialities. This means that women mostly reside in archetypal gender roles – as primary caregivers and managers of household affairs. At a broader level, men still maintain dominance in government and private sector institutions, as well as in other key organisations including the church.

For researchers, and in particular male researchers coming into the country with very different conceptualisations of gender discourses, the practice of individual masculinities in the field needs to be given greater importance in fieldwork planning and preparation. Very often, foreign male researchers enter Vanuatu with little understanding of the country's cultural norms, while also being disadvantaged by their lack of adeptness in the local language. Added to this is a poor understanding of salient issues around the performativity of individual masculinity, meaning that the propensity for researchers to understand the subtleties of the local context can impact the extent to which they can capture more representative understandings of societal attitudes.

The flipside can be an overzealousness to apply and insist on the researcher's own normative expectations – such as when it comes to gender empowerment and dismantling disproportionate patriarchal influence. While the intention to uphold a principled position may well have been virtuous, it may also have diminished good faith in male local actors. To what extent does a foreign, male researcher exercise the right to push back against local institutions and ways of doing things? To what extent is the insistence on promoting personal positions shaped under very different social and cultural frameworks? Furthermore, to what extent can a researcher's demonstration of masculinity stand apart from that which dominates the fieldwork study site and still achieve optimum outcomes? Such considerations are given limited attention during fieldwork training and preparation. Male researchers would do well to examine and understand the ways pre-existing masculinities are conceived of and applied at each of the study sites they work in, especially if they are diametrically opposed to their own.

Researcher 2

Background

As humans, we suffer from limits in our rational thinking brain's capacity to be aware of and to process the overwhelming environmental stimuli that our senses are exposed to. As a result, we attempt to simplify the world by generalising categories of stimuli. To further simplify this process, we draw upon generalisations and stereotypes that have been created, defined and accepted by the social groups that we identify with. The bifurcation of the world into a gendered male and female is one of the most fundamental of these generalisations (Bourdieu, 2001), deeply pervading all aspects of human existence and experience, from birth to death, and anthropomorphised onto non-human realms and inanimate objects, structures and activities of all types, including tourism destinations (Pritchard & Morgan, 2000).

At the same time, human societies have long recognised nuances in, and departures from, archetypal gender identities, although in only some cultural contexts have the more contrary of these variations been given equal recognition and status to the dominant gender norms. It is likely that in every human society (and perhaps non-human as well) there has always existed a tension between the stereotypical gendered norms and the individuals seeking self-expression that shatter those norms to some degree. In fact, it is easy to criticise stereotypical normative gender roles because so few people actually fit them in their entirety. Every individual person is an exception and there may be as many gender identifications as there are people on the planet.

The reason for this gendered diversity is because we are more than just gendered beings. We are also influenced by the many other ways in which we define ourselves as distinct individuals, which includes identifying with some generalised norms and disidentifying in opposition to others.

For me, my gender identity has been deeply entwined with my ethnic identity. I sometimes tell people that I am half Chinese and half German (both of my parents were immigrants), and that I am 100% American. I also might mention that the 50% of me that is German is on the outside, as seen in my facial features, whereas the 50% of me that is Chinese is on the inside, influencing what I most think I am. I grew up in a mostly white Californian suburb, but with access to Chinese and German immigrant relatives. However, I tended to blame all of my teenage inadequacies on my 'Chineseness'. My much stronger identity with my Chinese genes probably came from my being the eldest child (and a son) who identified with my Chinese father more than with my German mother. I also always felt that the west coast region of the United States was more Pacific Basin oriented than other parts of the country. However, it was not until after secondary school, and especially after going to Hong Kong to study for over two years, that I came to terms with and embraced my Chineseness.

While I identify myself as an American-born Chinese (ABC), I do not look Chinese so I am also a perennial outsider to that identity. I have compensated for this through my female relationships, which have always been with Chinese women, and in my research and writing which has mostly focused on East and Southeast Asia. While I speak some Cantonese (from Hong Kong) and Mandarin (from university study in the United States), it is not adequate to do research, so I am dependent on translators when I do fieldwork in Asia. Those translators are usually, although not always, female.

One characteristic that I identify, rightly or wrongly, as coming from my German heritage is my attitude towards women. My mother had what might be called a strong personality and I have always appreciated strong, intelligent and independent women. When given a choice in my fieldwork, I intentionally seek to promote the interests of female colleagues and students, to encourage them to be more self-confident and independent. Being a tourism researcher, opportunities to do this are plentiful because the lower academic echelons of tourism studies tend to be dominated by female academics and students across the regions of Asia that I work in.

I feel very uncomfortable in all-male research and academic leadership settings, which I have most often encountered in Hong Kong and South Korea and, to some degree, in Japan, but less so in Southeast Asia. Maybe that is because such settings remind me of secondary school, and maybe I am afraid of being held to the gender norms and behaviours of those days which I have long since rejected – especially with regard to the sexual objectification of women. Thus, while in some ways I strongly identify with my gender and masculinity (for example, in terms of leadership and independence), in other ways I strongly do not identify with my gender and masculinity (for example, in terms of male bonding, female stereotyping and gender superiority). Being an academic has allowed me a greater degree of flexibility in shaping my maleness than perhaps some other professions might, although male academics are also gender stereotyped to a large degree, as I suggest below.

Researcher 2: In the field

I have had the great fortune to be able to travel, live, study, consult, lecture and conduct research in Asia for over 40 years. I know East and Southeast Asia best, but have experienced most other parts of the Asia-Pacific realm as well. Through these decades of experience, I have always seen myself as an outsider, although sometimes more so than at other times. This perspective has helped to keep me sensitive and self-reflexive to the roles that I am playing as a male American scholar in Asia. When conversations are taking place in a language that I cannot comprehend, for example, my mind often seeks to understand why I am here and how I am relating to, learning from and contributing to this context as an outsider. Often this includes thoughts of both the advantages and the disadvantages that I have in such a setting. Sometimes this includes thoughts on the gendered nature of a given situation, which may be overly male or overly female – both of which can trigger concerns.

However, gender is not my only area of reflexive concern. It is one of multiple identities that give me advantages and disadvantages when I do fieldwork in Asia. In addition to my maleness, other identities that often set me in a superior positionality compared to those I am working with include:

(1) my age (I am usually older than those I work with and therefore more deserving of Asian respect);
(2) my Euro-American (white) facial features (giving me deference by some over my Asian colleagues);

(3) my Asian ethnicity, along with my basic Chinese speaking abilities (which come as a surprise and often create a connection, especially with the Chinese I work with or encounter);
(4) my English language speaking and writing abilities (which Asian colleagues are envious of, and which they often see as a resource or utility);
(5) my developed country (US) nationality and residence (which gives me an advantaged income and access to scholarly resources); and
(6) my reputation as a scholar (which I consider over-exaggerated and mostly reflecting the fact that I have been doing this work for a longer time than many others).

On the other hand, I also have some disadvantages, which can somewhat counter-balance my clear advantages. These include:

(1) my lack of fluency in any of the major languages in Asia (which limits my ability to communicate and understand what is going on);
(2) my sobriety, which prevents bonding over alcohol (I stopped taking mind-altering substances when I was about 21 years old); and
(3) my innate introversion and shyness (something I associate with Chineseness, and which further limits my sociability and connections with people).

Both the advantages and disadvantages that are embedded in my fieldwork experiences bring biases to the research process by introducing preconscious predispositions. These biases can only be approached through the intentional reflection and questioning of my research role, along with that of other participants in the research setting. This, however, is seldom approached in a structured manner, but rather occurs only when such biases become too apparent to ignore.

Upon reflection on my research and broader academic career in Asia, I believe the following four multidimensional roles have typified the breadth of my experiences. (These are written in such a way as to not identify individuals I have worked with.)

(1) *The student and the teacher.* The first decade or so of my academic career involved occasional travel to Asia, initially as a PhD student and then as a temporary lecturer. My experiences were mostly urban, in Hong Kong and Singapore, and as best as I can recall I was largely oblivious to gender issues. Some of this included travel with my spouse and children.
(2) *International resource sponsor.* About 15 years after my PhD, I became more involved in the International Geographical Union's Tourism Commission, and started the *Tourism Geographies* journal. From that background I also started to assist in organising international conferences, most of which have been in some part of Asia. I have been primarily responsible for managing international

participants for these conferences. Over the years, this has involved some combination of promoting the event internationally, creating conference websites in English, collecting foreign participant registration fees, identifying and inviting international keynote speakers and assisting with post-conference publications. Publishers have specifically told me that they only agreed to a post-conference book for some of the conferences because I was one of the editors. In return, I usually get a free trip to participate in the conference. I am clearly leveraging my positionality as a Western, developed world scholar and native English speaker with access to resources that are out of easy reach for academics in developing countries. My maleness enhances my access to these resources, which even for a woman with similar attributes could be more challenging. From the perspective of my local colleagues, however, my positionality is a valuable resource which they can leverage to achieve their own personal goals.

(3) *The Foreign Research Partner.* I started being invited to be a research partner on proposals written by Asian colleagues starting about 25 years or so after my PhD, but only a few of these were successfully funded. For funded projects, I work closely with senior academic scholars, although they are usually not quite as old as me. Most of these scholars are male, although a couple have been respected female professors. These research projects involved more in-depth participation and site visits on my part, and often included intellectually stimulating debates over the conceptualisation of the research, the methodology and the results. In general, my role tends to focus more on the theoretical context of the research, whereas my collaborators contribute local knowledge and expertise. While I am still in an advantaged position, the relative difference between local scholars and myself is considerably less, and in some areas they are positionally advantaged over myself. I believe that my mixed ethnicity, being both an American and an ethnic Chinese, contributes significantly to my success in this role by balancing my social bonding with Asia and my being a representative of the so-called 'developed Western world'. Maleness is a significant advantage when working with male Asian partners who are made more comfortable. My role as a senior male mentor, however, places me in an advantageous position among both male and female collaborators.

(4) *The senior male mentor.* This is a role that I started to take on later in my academic career (about 25–30 years after my PhD) and is prominent today as I enter retirement (see Figure 7.4). In this role I am invited to be an external advisor for research projects that are primarily undertaken by a younger, usually female, scholar. I am introduced to the research site and issues in the same way as a tourist might be on a personalised tour of a destination. The site visit is relatively brief and I am expected to provide insights and perspectives to make the

Figure 7.4 The senior and foreign male mentor
Source: Lew (2019).

research more robust and publishable. It is as if I am serving as a peer reviewer of a research proposal or a paper for a journal. I provide comments, and these are apparently appreciated. But I often question just how significant my comments might be, due to the time limitations that add to the biases of my positionality (listed above). In this context I am clearly levering my age, my foreign identity, my scholarly reputation and other elements of my advantaged status. Maleness, however, seems to underlie all of these elements. While a female scholar may serve a senior mentor role, my experience and knowledge of Asia suggests that this is rare, or at least at a much lower rate than among male scholars. While I am sure there are exceptions, it is generally safe to say that most male academics in Asia are also more inclined to accept guidance from an older, foreign male than from a female scholar of any age and origin.

Researcher 2: Implications

As academic scholars, we undertake research and fieldwork because it is intellectually stimulating and because we want to share what we learn with our peers. Our positionality in that research, however, must be recognised to fully understand the value and quality of what we are learning and sharing with the world. Unfortunately, this is probably seldom done, especially by male researchers who can too easily ignore their taken-for-granted dominant status. After all, it feels good to be advantaged, praised and put on a pedestal, so there is little incentive to question this.

A further disincentive for self-reflection is the effort required in disentangling the complexity of roles that all researchers bring into the field. Gender is not the only role that shapes the advantages and disadvantages that so easily bias our research. However, gender may be the most fundamental, being the identity that we have been most closely connected to since the earliest moments of our lives. An awareness of the major types of roles that academics may encounter in their fieldwork can help to simplify an understanding of the gendered contexts that we encounter. The four role types identified above are very specific to my personal experience but may provide a guide for developing one's own self-reflexive model of cross-cultural fieldwork experiences.

On the other hand, few experiences offer the same potential for deep, inner reflection and insight into ourselves as embedded, cross-cultural fieldwork does. This may be the ultimate reason why many scholars seek out such experiences, because they offer not only professional growth, but also personal growth. We are completely removed from our mundane, taken-for-granted existence and repositioned in a liminal setting that offers opportunities for expanded creativity and new relationships with ourselves and our world (Lew, 2020).

Conclusion

In this presentation of two distinct illustrations of the performance of masculinities in the field, we have tried to demonstrate the abiding influence that cross-cultural influences have had on the practical application of academic research. In both cases, our respective backgrounds have had an overarching influence on our respective research practice, and have shaped the social construction of how we conceive of ourselves as researchers. The questions posed at the outset concerned the extent to which our performance of masculinity in field research has been negotiated and shaped by cross-cultural contexts. Our stories, in response to those questions, draw from our individual standpoints insofar as ethnicity, nationality and age, among other things, are concerned, as well as the sites in which much of our work has been undertaken.

Several emergent themes are obvious, with the overarching leitmotif being that the performance of masculinity in both cases has been socially constructed. The countries we live in, the universities we ply our trade with, our language competence (both English and other), study sites we have been drawn to (in the Asia Pacific), our familial and multi-ethnic make up, and of course the fact that we are both male researchers, underline our practice of masculinities when in the field. At the risk of oversimplifying matters, the questions concerning how to be an 'academic man', and what kind of a man that would be, tend to resonate in discussions about contemporary masculinity as reflected in some of the contributions to this volume.

In our case, our upbringing and exposure to diversity in our childhood contexts has helped shape a tolerance to difference and an abiding openness to socially progressive perspectives. The question concerning the extent to which the male role models and father figures in our lives have influenced our dispositions with regard to how we relate to female counterparts is a pertinent one. Much has been said about how young men pivot and perform according to the lessons learned from the influential males in their lives and in our cases this is a relevant point. Conversely, the other related question with potential for deeper investigation is how the prominent female figures in our lives have shaped our attitudes towards women. In both our cases, central to our upbringing is an absence, or at least a modification, of popular gendered stereotypes, where both our parents coexisted in a context underlined by equality, in which roles that are typically gendered were not so clearly defined.

That we both have a research practice steeped in the Asia Pacific and maintain pan-Asian and Eurasian identities gives us both etic and emic perspectives that are less available to researchers who do not share a similar make-up. This is very often advantageous in fieldwork situations in the Asia Pacific where we also maintain a privileged status as foreigners and academics. Furthermore, our distinct gender roles reinforce our privileges in ways that local-based researchers in the countries we work in may not have access to. Very often this privilege intersects with how we exercise our masculinities, particularly in situations where gender disparities are predominant. This is especially evident where our academic leanings in the humanities find us more often than not working alongside a diversity of women as colleagues or informants. What to do in cases where our own sense of gender relations clashes or sits awkwardly alongside fieldwork situations remains a continuous and reflexive undertaking.

The central question then remains: how have all the contributing influences underscored the performance of masculinity in our research practice? Overall, it seems that the socially constructed norms that were established in our formative childhood and adolescent years continue to resonate in and cross over into our personal and professional identities. As researchers, we take our assemblage of attitudes, opinions and norms into the field, allowing and embracing diversity as a personal reference point. This arms us with the ability to adapt and respond to the multiplicity of situations where negotiating cross-cultural contexts adroitly for productive effect is integral.

Researchers who have not been exposed to diversity and difference may find such situations uncomfortable and awkward, which can be a constraint to negotiating and moderating how their own masculinities align with those of their research sites and participants, especially where they are outside their home base. How then to prepare male researchers with little experience of diversity and difference becomes a key question. After all, notions of masculinity and how this is enacted in research practice can determine the success or failure of fieldwork endeavours where the gulf between a researcher's cultural constructs and those of their participants is considerable. What then are our obligations as researchers if we stumble upon ethical and moral dilemmas related to the performance of masculinity that are at odds with what we are generally used to?

Finally, the implications of the dual vignettes presented, and the subsequent analysis, suggest that adroitly negotiating and performing masculinity on a global and culturally diverse stage starts with an embracement and embodiment of diversity in the first place. Can this be learned or does it become second nature depending on the nature of the individual's background? And, would an aversion to or fear of cultural diversity render a male researcher's ability to foster favourable and productive encounters null and void? Clearly such generalisations are left wide open to scrutiny. However, the increasing drive for gender parity and complementarity will demand that male researchers hone their research practice to ensure that their conceptions and performance of masculinity are continually reset in the #metoo era rather than relying on default settings. After all, behaviours and attitudes that were once tolerated and considered benign are now subject to a more vigilant and critical scrutiny. This means that the barbarians at the gate can no longer ride roughshod over gender equality concerns and instead will have little choice but to adapt and reconfigure to find the female and male gifts within all of us.

References

Ali, B.N.K. (2005) Quong Tart and early Chinese businesses in Fiji. *Journal of Pacific Studies* 28 (1), 78.
Bennett, M. and Sani, F. (eds) (2004) *The Development of the Social Self*. Hove and New York: Psychology Press.
Botterill, D. (2003) An autoethnographic narrative on tourism research epistemologies. *Loisir et société/Society and Leisure* 26 (1), 97–110.
Bourdieu, P. (2001) *Masculine Domination*. Stanford, CA: Stanford University Press.
Butler, J. (1990) *Gender Trouble: Feminism and the Subversion of Identity*. London: Routledge.

Chang, H. (2016) *Autoethnography as Method*. London: Routledge.
Chattier, P. (2015) Women in the House (of Parliament) in Fiji: What's gender got to do with it? *The Round Table* 104 (2), 177–188.
Clatterbaugh, K. (1998) What is problematic about masculinities? *Men and masculinities* 1 (1), 24-45.
Cornwall, A. and Lindisfarne, N. (eds) (2016) *Dislocating Masculinity: Comparative Ethnographies*. Abingdon: Routledge.
Crowley, T. (1995) The national drink and the national language in Vanuatu. *Journal of the Polynesian Society* 104 (1), 7–22.
Everingham, P. (2016) Hopeful possibilities in spaces of 'the-not-yet-become': Relational encounters in volunteer tourism. *Tourism Geographies* 18 (5), 520–538. doi:10.1080/14616688.2016.1220974
Haig, M. (2016) The Descent of Man by Grayson Perry review – a man's man is yesterday's hero. *The Guardian*, 23 October. See https://www.theguardian.com/books/2016/oct/23/descent-of-man-masculinity-grayson-perry-review-a-mans-man-is-yesterdays-hero-gender-role (accessed 1 February 2019).
Hermann, E. and Kempf, W. (2005) Introduction to relations in multicultural Fiji: The dynamics of articulations, transformations and positionings. *Oceania* 75 (4), 309–324.
Huffer, E. and Molisa, G. (1999) Governance in Vanuatu: In search of the Nakamal way. Discussion Paper No. 99/4, Research School of Pacific and Asian Studies. Canberra: Australian National University (ANU).
Kagan, D. (2017) Legacy masculinity: Grayson Perry's The Descent of Man. *The Lifted Brow* 35, 29–32 (online).
Kimmel, M. (n.d.) Raise your son to be a good man, not a 'real' man. *The Cut*. See https://www.thecut.com/2018/03/teaching-our-sons-to-be-good-men.html (accessed 1 February 2019).
Leckie, J. (2005) The complexities of women's agency in Fiji. In B.S.A. Yeoh, P. Teo and S. Huang (eds) *Gender Politics in the Asia-Pacific Region* (pp. 168–192). Abingdon: Routledge.
Lew, A.A. (2020) The global conscious path to sustainable tourism: A perspective paper. *Tourism Review* 75 (1), 69–75.
Mann, C.W. (1939) A test of general ability in Fiji. *Pedagogical Seminary and Journal of Genetic Psychology* 54 (2), 435–454.
Mutua, A.D. (2013) Multidimensionality is to masculinities what intersectionality is to feminism. *Nevada Law Journal* 13 (2), Art. 4.
Noy, C. (2008) The poetics of tourist experience: An autoethnography of a family trip to Eilat. *Journal of Tourism and Cultural Change* 5 (3), 141–157.
Patterson, M. and Elliott, R. (2002) Negotiating masculinities: Advertising and the inversion of the male gaze. *Consumption, Markets and Culture* 5 (3), 231–249.
Perry, G. (2017) *The Descent of Man*. New York: Penguin.
Pritchard, A. and Morgan, N. (2000) Privileging the male gaze: Gendered tourism landscapes. *Annals of Tourism Research* 27, 884–905.
Sahlins, M.D. (1963) Poor man, rich man, big-man, chief: Political types in Melanesia and Polynesia. *Comparative Studies in Society and History* 5 (3), 285–303.
Scarles, C. (2010) Where words fail, visuals ignite: Opportunities for visual autoethnography in tourism research. *Annals of Tourism Research* 37 (4), 905–926.
Taylor, J.P. (2008) Changing Pacific masculinities: The 'problem' of men. *Australian Journal of Anthropology* 19 (2), 125–135.
Vanderbeck, R.M. (2005) Masculinities and fieldwork: Widening the discussion. *Gender, Place & Culture* 12 (4), 387–402.

8 Gender, Identity and Discomfort: Negotiating Self and Context in the Field

Dominic Lapointe

I'm a man.
I'm a white man.
I am privileged.

Addressing the issue of gender in field research was an introspective journey for me. My first thought upon receiving the invitation to submit an abstract for the book was: 'Well, I've never touched this topic in my research.' At the same time, I was reviewing a chapter that used autoethnography to analyse the influence of standpoint in research construction (Benjamin *et al*., 2019). The chapter focused primarily on race and class, and despite the female gender of two of the three authors, gender was only briefly mentioned. This lack of discussion was a clear expression of gender as an invisible methodological category (Bell *et al*., 1993; Monjaret & Pugeault, 2014). The absence of gender as an issue in methods and fieldwork analysis is a recurring theme; this motivated me to launch another round of autoethnography and revisit my field notes.

I will first present an autoethnography as a way of inquiry and self-reflection. Afterwards, I will move back and forth between the conceptual and the personal, looking at masculinity, positionality, biographies and fieldwork. I will close with an inconclusive conclusion that explores the idea of discomfort in research.

Autoethnography

Using autoethnography to dive into personal aspects of conducting research is a strategy to situate knowledge (Haraway, 1988) within its contexts of gender, social class, ethnicity, etc. Researchers over their life course develop a posture and a relationship with their subject (Briscoe &

Khalifa, 2015). The role of positionality in motivations to work with certain topics, to build research as well as to develop a posture, appears a relevant reflection to establish a critical self-reflective look at research. When researchers use standpoint theory:

> They engage in the practice of science critique aimed at countering the erasure of the situated interests that shape the sciences, and they are particularly attentive to the role played by dominant systems of knowledge in reproducing and amplifying systems of oppression. (Wylie & Sismondo, 2015: 328)

A critical perspective involves researchers thinking about their own position in society, situating their voice, and looking at research in an embodied, reflective and situated way, without falling into relativism (Botterill, 2007). At the same time, this perspective influences our gaze and the way we experience daily life, power, alienation and vulnerability. Moreover, deconstruction, via a self-reflexive approach, addresses the impossibility of abstracting ourselves from the data-collection process, especially during the field research phase (Kleinman, 1991). This phase, which is more a question of interaction than of observation, implies the researcher's subjectivity as an active subject in research, hence the importance of making this aspect of research visible. 'We must write about why we have the setting, who we are at the moment, and how our identity affects our reactions to the setting and its participants' (Kleinman, 1991: 194). As such, the works of Briscoe and Khalifa (2015) are exemplary and directly inspired my choice to use autoethnography in this chapter.

The positionality of researchers also influences their interactions with research subjects. Kobayashi (2001) situates her position as an Asian woman as an influential factor in her reading of the data and in her relations with the research subjects. She considered social inequalities to be an integral part of her research, and called for disclosure to foster an ethical and critical discussion on its role in research processes and methods. Porter and Schänzel (2018) describe gender positionality as an important differentiating factor in conducting field research, in the differences it generates in access to informants, in stereotypes faced by researchers and also in how empirical data is read by situated female voices. Thus, it would be difficult, if not impossible, to generate an understanding of the oppression and subalternity of the unique universalising view of the dominant (Haraway, 1988), which is mostly the view of affluent white males.

Again, following the work of Briscoe and Khalifa (2015), I used autoethnography as a source and assumed their rhizomatic posture of knowledge (Deleuze & Guattari, 1980). This approach is based on a non-hierarchical, multiple and shifting vision of knowledge, where different entities (in this case, my experience and my research) combine and interact in complex and unexpected ways, sometimes forming a stable stratified assemblage. The strata are stable combinations of content and expression, where content is made of substance (what is) and form (organisations of substances), and expressions also have substance (the phenomenon) and form (institutions). Content and expression are interrelated within the strata. Knowledge is created by following threads of the rhizome in a disentanglement process of those complex knots (Lowenhaupt Tsing, 2015) within the strata.

Who Am I? Negotiating the Personal and the Political While Doing Research

Stating who we are and how we construct ourselves in, with and through context at the intersection of agency, structures, meaning and theories is an important part of doing fieldwork. Going further with the Deleuzian constructs of stratification and rhizome, the stratification from my standpoint represents stabilised interconnected layers of meaning about who I think I am (expression) and how I think I act (content) according to those layers, which can potentially erode, explode or otherwise be shocked into different combinations. The strata are masculinity (substance) and social class with ambiguous colonial background (forms). I will go through these strata via autoethnography and literature, and then I will review how the conceptualisation of my fieldwork can be read through those strata. I will illustrate with an episode of my PhD fieldwork and close the chapter with a reflection about discomfort as a field attitude to cultivate instead of trying to avoid it.

Masculinity as Dominant Hegemonic Construct

As Butler (1990) states, gender is a construct negotiated through an array of norms and conventions. As far back as I can recall, my identity has been associated with the masculine side of the gender spectrum. Growing up in a blue-collar working-class environment I was confronted early on with the codes of masculinity in this particular class environment, the face of hegemonic masculinity (Connell & Messerschmidt, 2005) and my difficulties with corresponding to its standards. These standards often have close ties to what I usually call 'the hockey locker room brotherhood' with its expectations of strength, arrogance and bonding via comments about the other spectrum of gender, femininity. As Vanderbeck (2005) states:

> Hegemonic masculinities are culturally sanctioned ways of being male which are associated with power, such as physical strength (hetero)sexual potency, authoritativeness and self-assurance. (...) Non-hegemonic masculinities, in contrast, are 'inferior' ways of being male which do not conform with hegemonic ideologies. Certain styles of self-presentation, for example, can be denigrated as effeminate or otherwise deficient. Failure

> to respond effectively to physical or verbal challenges from other males can place a man in a subordinate position in the hierarchies that men construct amongst themselves. (Vanderbeck, 2005: 390–391)

All my life, my love of books, my way of caring, of crying, of feeling was associated with being not close enough to what hegemonic masculinity was calling for, resulting in other men expressing doubt about my sexual preferences. As a male kid, teen and young adult, in my surroundings I felt an ambiguity, divided by the sense of pride about my good marks and capacity to learn things and the uneasy sense of slight shame that some of my subpar 'male' skills were creating around me. I still can't cut a piece of timber and I still prefer cooking with my mom and aunts at Thanksgiving over sipping beer in the garage around the car and tools with the men in my family. The identity assignation of hegemonic masculinity was and still is something I negotiate, between social acceptance and a personal sense of who I am in a blurred gender/sexuality spectrum.

University was both a relief and a trap. A relief, because I felt I finally fit in somewhere. A trap, because being a first-generation university student, it took me quite a while to figure out that I was reproducing another form of hegemony, that of the white male scholar (Armato, 2013). This form of masculinity was asserting its hegemony in a different way from the one I was used to in my working-class environment, placing me in an identity conflict. Indeed, the masculinity of my social class of origin, which I had always been uncomfortable with or even excluded from, was dismissed as inferior to the one expressed in academia. Yet the class aspect of my identity, which can easily be spotted by my use of language and relationship to dress codes, alcohol, etc., was also expressed as lesser. As Armato (2013) describes it:

> We cultivated a sense of distinction between ourselves, as academic men, and the world beyond the academic context; this was primarily accomplished by our allusion to the men 'out there' who remain unnamed, but who I would argue were, in our consciousness, working-class men and men of colour. (Armato, 2013: 586)

I found myself in a situation where I was at the same time the academic man and the man 'out there'.

Social Class and the Relationship to Economy and Development

I was born in a working-class neighbourhood in Quebec City, between two factories and a highway. My parents were barely out of their teens. Even today, I feel the imprint of where I come from, the physical place and, especially, the social class. Long before encountering the writings of Marx, Engels, Lefebvre and Bourdieu, I understood, without knowing the words or the concepts, the existence of social class and alienation, and the importance of relationships with others and with capital (money). There was a constant undercurrent of insidious forces constraining and dictating the place, and your place in it. Alcohol was associated with great-uncles and great-aunts; it was omnipresent in everyday life, good times and bad. *Maudite boisson*[1] – 'Goddamned drink' – my grandmother would always say. She was the only one of her generation who didn't drown out her alienation with alcohol. The stories of this time still percolate in my current identity – stories of a Quebec that is fading from memory, due to deindustrialisation and economic restructuring, but one that marked the generation of my grandparents. A Quebec known for crowded family homes, factory bosses who only spoke English, a French full of *joual*[2] – its expressions still fill my language every day. There were stories of the Dominion Textile and the Anglo-Pulp Paper where people lied about their age to start working. I remember the feeling when the Dominion closed in 1987. I was 13. I remember worrying, 'What are the people losing their jobs going to do?' Resignation, the loss of a landmark, the end of an era. Misunderstanding, astonishment, but also resentment, that's what I heard when faced with a situation that was a tragedy for many. Even though nobody in my family worked there, the conversations about it were recurring, sharpening my awareness and my desire to understand what I would later learn to name as class structures.

An Ambiguous Colonial History

In 1994, enrolled on a sociology course at the University of Saskatchewan, I was confronted with the political dimension of my French accent in an English-speaking world on a daily basis. This was a period when Quebec was on its way to a referendum on autonomy. Every day, there were small comments, or questions like 'What does Quebec want?' The all too classic comment on the battle of the Plain of Abraham, 'You lost the battle, you should move on …'. Just who was the 'you' they were referring to? I experienced a job interview being cut short as soon as I said that my accent was not from France but from Quebec. There was a permanent feeling of being a stranger in the country that I was born in, Canada.

Intellectually, I was also confronted at this time by an important contradiction when a sociology course used an analysis of the Quiet Revolution (Shibutani, 1986) to illustrate social change. It came as an electroshock to read (in French) about the historical alienation of French Canadians from Anglo-Saxon capitalism, while following a sociology course (in English) about the Quiet Revolution, a period of emancipation for French Canadians living in Quebec, topped with daily reminders of my language, my roots and the political malaise that they embodied under the big blue sky of Saskatchewan. My positionality of the majority in Quebec became a carrier of subalternity outside the province. This majority/minority, dominant/subaltern contradiction was part of me, part of where I'm from. I live with an ambiguous colonial past, as both a coloniser-settler figure expressed through the French subsidiary of the Canadian crown, but also as a colonised figure present in the social class I was born in.

> *Elle contient également une figure de colonisé, plus ancienne, moins saisissable, moins aimée: une population native, illettrée, attachée juridiquement à sa subsistance dans les seigneuries des deux rives du Saint-Laurent ou lancée à l'aventure sur les routes commerciales antiques. Cette population est soumise aux aléas de l'histoire européenne, passant du joug français au joug britannique sans jamais y prendre part, (…). Cette population sera à partir du 19ᵉ siècle graduellement arrachée à la terre ou au nomadisme et captée par le capitalisme impérial et l'économie extractiviste contemporaine sous la forme d'une forme de travail.*[3] (Giroux, 2019: 39)

Those two figures were echoing the two masculinities I negotiate with: two conflicting positions, two components of an imposed social hierarchy that I must account for.

Being in the Field

My field research focuses on issues of development, community and justice in peripheral areas where tourism is brought in as the solution for development (Jeannite & Lapointe, 2016; Lapointe & Gagnon, 2009). Revisiting my notes made me realise the extent to which regional development and planning is a gendered sector. Although it cannot be statistically generalised from my research, I can identify some recurring patterns of gendered role distributions. For the most part, politicians, government officials and entrepreneurs/developers are men, and women tend to fill social and cultural roles. Right from the beginning of my PhD, my female supervisor made me aware of the importance of being sensitive to gender differences in fieldwork and of obtaining the best possible parity between men and women in my interviewees. This also reproduced a fracture between economic development and social/cultural development discourses where the diversification polarised my samples into two major quadrants: male economic versus female social/cultural. This was not surprising and in retrospect I should have tried harder to obtain higher diversity in the two other quadrants.

Projections, Witness and Defiance

In the field I tend to neutralise myself and re-engage critically with the materials afterwards. This strategy creates a projection from the participant. Reviewing my notes has made this all the more obvious, and it is an aspect that definitely deserves more scrutiny in my future research projects. Indeed, my transcripts reveal differences between men and women in the interview process, which is also part of the economic/social divide of the informers. Men in dominant economic positions tended to project that dominant hegemonic masculinity and economic perspective on me. As an expert they asked me to witness their accomplishments, but they also sought my approval: there was a projection of their positionality on me. They set the interviews in their offices, behind closed doors and maintained confidentiality as one of the principles of the interview. They used a lot of sentences like 'As you may know...', 'I know you understand...', 'Like you might have seen somewhere else ...', all the while trying to create male bonding along the lines of hegemonic positionality. They sometimes even went as far as to offer drinks during the interview process; they too may have had a subaltern side to their positionality.

On the other hand, female informers rarely offered me interviews behind fully closed doors, finding in-between spaces where they could be recorded away from other ears without being in a fully enclosed context that could be misinterpreted as intimacy. They didn't look for bonding in answering my questions, and talked more about their organisation's actions than their own individual accomplishments. Trying to be neutral and not acting as a hegemonic male or as my caring, sensitive self involved having different projections on me as a male researcher. First of all, I had to meet my female informers more than once in order to actually have a recorded interview; the trust-building process took longer. In my research in Quebec there was often a generation gap, as an aging population meant doing interviews with people closer to my grandparents' age than mine; in Guadeloupe there was also a racial difference as almost all of my informers were Afro-descendants. Those differences were also at play, along with the gender difference, in the trust-building necessary to complete the interviewing process. Furthermore, in Guadalupe I always made sure to be clear about the ambiguous colonial past of my positionality, marking the difference between the French-Canadian and the French from the Métropole,[4] but I tended to neutralise the gender difference, not declaring my own negotiations with masculinities and its hegemonic side.

Standing in Front of the Board

In 2007 I was completing my fieldwork in Côte-sous-le-Vent on Basse-Terre Island in Guadeloupe. My positionality in this field was one of a white man, not from France but from a group with a colonial past, both as colonisers and as colonised. This positionality and its impact on my interviews was discussed in my thesis (Lapointe, 2011), but was particularly expressed during a session reporting to the Board of the *Parc national de la Guadeloupe* (Guadeloupe National Park) at the end of the fieldwork. At that moment, I was invited to speak to the entire park team. The directors attending the meeting were white metropolitan (mainland France) civil servants with the exception of those from the sustainable development department, all of whom were Afro-descendants. Throughout my presentation, I felt stuck in my contradictions where I was simultaneously subordinate, in my language and my class origins, and privileged by my whiteness and my masculinity. The environmental and social justice perspective and discourses I presented on the relationship between the park and the neighbouring communities disturbed a dominant and subalternising institutional view that served to justify asymmetrical relations with these neighbouring communities. These relations favoured those from mainland France over Afro-descendants, who are also French citizens.[5] However, my scientific language also conveyed a relationship of domination. The questions during the cold and uncomfortable Q&A period came mainly from the metropolitan directors, and implied that as a foreigner I did not quite understand everything, which activated a slight imposter syndrome in my young researcher's mind.

Once it was over, as the room was clearing out, a female Afro-descendent park fieldworker glanced at me and said, *Merci*. I understood, a few days later in another meeting before my departure, that I had, in my words as a scientist, expressed what this field officer had been repeating for several years. The white man that I am was allowed to speak, despite my youth, my Quebec otherness in this persistently colonial context and the doubts expressed by the senior leaders. After only three months, I had the right to express what I had seen and heard and the team deemed me worth listening to. The fact that I was white, educated and a man most certainly played a part in this dynamic. This episode illustrates the intersection between a privileged situation and subaltern class and cultural origins.

> Out of the framework of multiple masculinities and femininities emerges a portrayal of a hierarchical structure of gender relations which includes inequalities between men and women, among men, and among women. Of importance here is that to speak of these complex sets of relations as a structure does not mean that it is a static system. (Armato, 2013: 580)

In the situation I describe, those different layers of identity and positionality were at play. As a white academic man, I was allowed to speak and present my argument, but at the same time I was reminded that I was young and from a peripheral geography and a lower class by the park's upper management. Even if I could rely on a rigorous method, interviews and a full three months of continuous interactions and note taking, the management challenged my explanations with stereotypical explanations about race and behaviour to minimise my findings. At the same time, for the Afro-descendent fieldworker who spoke to me, I was still a white academic man with all the attached privileges. Her simple 'Thank you' carried many meanings, confirming my privilege to speak and be heard after only three months and confirming that my findings made sense from her standpoint, which spoke to my working-class origins, but also bringing a form of the white (and masculine) saviour narrative to the situation. Exposing injustice to the park's management team was also a form of appropriation of the voices of those not heard. This complex entanglement was creating a strong discomfort; it brought up feeling of restlessness, anger, pride and sadness in me. This discomfort was like being tied in a knot. Pulling the string to get out of the feeling – no matter the direction – only made it tighter. I could not get away from any part of it, whether I liked it or not.

Inconclusive Final Remarks of an Ongoing Thought

Concluding this chapter seems inadequate. How can we conclude something that is ongoing? We constantly negotiate who we are, whether we are aware of it or not. No matter how well stratified it is, the rhizome can always go a different way and reorganise itself and restratify with new combinations. In this ever-unfolding and refolding process, we tend to avoid discomfort, looking for comforting signs of recognition and accomplishment. Looking at the stream of thoughts I layered in this text, I would argue that in the field, discomfort can be a tool of inquiry to be accepted and welcomed instead of solved. Indeed, the act of having a self-reflective space to write and communicate about my own self, my own dilemmas and contradictions, my pride and doubt, is a privilege in itself.

> Researchers seeking to practice reflexivities of discomfort do not dismiss the importance of examining issues of power and ethics but recognise that reflexivity is inextricably linked to power and privilege that cannot be easily or comfortably erased. (Burdick & Sandlin, 2010: 354)

I am still a white and privileged man, and no matter how aware and well-intentioned I want to be, this should not create a comfortable space of inquiry where I can say: I'm not one of them. In the example I described, I could not get away from the masculine identity projected on me, no matter how I negotiated it; nor could I shirk off my whiteness, my working-class background and or the particular colonial context of my identity. I believe – and it is the accurate word, even if it is a word coined as not scientific – I believe that discomfort needs to be embraced as a tool of disentanglement through complex identities and contexts, a way to keep in view that even when we fully acknowledge injustice and unequal power relationships, we might insidiously keep reproducing them. This calls for scrutiny and constant re-evaluation of how we act and think.

Notes

(1) This expression from Quebec was commonly used in working-class environments to complain about alcohol, mostly its social impacts.
(2) *Joual* refers to the specific slang of the lower-class French-speaking people of the St Lawrence River Valley.

(3) I intentionally left the quote in French; here is my own translation: 'It also contains a colonised, older, harder to grasp, less beloved figure: a native, illiterate population, legally bound to subsistence livelihood in the seigneuries on both shores of the St Lawrence or looking for adventure on ancient trade routes. This population is subject to the vagaries of European history, passing from the French yoke to the British yoke without ever taking part, (...). Starting in the 19th century, this population would be gradually torn from the land or nomadism from the 19th century and captured by imperial capitalism and the contemporary extractivist economy in the form of a workforce' (Giroux, 2019: 39).

(4) La Métropole is the word used in the DOM-TOMs (French Overseas Departments and Territories) to refer to the European France.

(5) For more on this work, see Lapointe and Gagnon (2009). For more on Guadeloupe's particular sociopolitical situation, refer to Rauzduel, (1998).

References

Armato, M. (2013) Wolves in sheep's clothing: Men's enlightened sexism and hegemonic masculinity in academia. *Women's Studies* 42 (5), 578–598.

Bell, D., Caplan, P. and Karim, W.J. (1993) *Gendered Fields: Women, Men and Ethnography*. London: Routledge.

Benjamin, C., Cosaque, C. and Lapointe, D. (2019) Positionnalité et recherche critique: Diversité de construction d'un même objet et émergence de la critique. In H. Bélanger and D. Lapointe (eds) *Perspectives critiques et analyse territoriale: Applications urbaines et régionales* (pp. 168–187). Québec: PUQ.

Botterill, D. (2007) A realist critique of the situated voice in tourism studies. In I. Ateljevic, A. Pritchard and N. Morgan (eds) *The Critical Turn in Tourism Studies* (pp. 143–152). Abingdon: Routledge.

Briscoe, F.M. and Khalifa, M.A. (eds) (2015) *Becoming Critical: The Emergence of Social Justice Scholars*. Albany: SUNY Press.

Burdick, J. and Sandlin, J.A. (2010) Inquiry as answerability: Toward a methodology of discomfort in researching critical public pedagogies. *Qualitative Inquiry* 16 (5), 349–360.

Butler, J. (1990) *Gender Trouble: Feminism and the Subversion of Identity*. London: Routledge.

Connell, R.W. and Messerschmidt, J.W. (2005) Hegemonic masculinity: Rethinking the concept. *Gender & Society* 19 (6), 829–859.

Deleuze, G. and Guattari, F. (1980) *Milles plateaux*. Paris: Les éditions de minuit.

Giroux, D. (2019) *Parler en Amérique: Oralité, colonialisme, territoire*. Montréal: Mémoire d'encrier.

Haraway, D. (1988) Situated knowledges: The science question in feminism and the privilege of partial perspective. *Feminist Studies* 14 (3), 575–599.

Jeannite, S. and Lapointe, D. (2016) La production de l'espace touristique de l'Île-à-Vache (Haïti): Illustration du processus de développement géographique inégal. *Études Caribéennes*, April, 33–34.

Kleinman, S. (1991) Field-workers' feelings: What we feel, who we are, how we analyze. In W.B. Shaffir and R.A. Stebbins (eds) *Experiencing Fieldwork: An Inside View of Qualitative Research* (pp. 184–195). London: Sage.

Kobayashi, A. (2001) Negotiating the personal and the political in critical qualitative research. In M. Limb and C. Dwyer (eds) *Qualitative Methodologies for Geographers: Issues and Debates* (pp. 55–72). Oxford: Oxford University Press.

Lapointe, D. (2011) Conservation, aires protégées et écotourisme des enjeux de justice environnementale pour les communautés voisines des parcs? Doctoral dissertation, Université du Québec à Rimouski.

Lapointe, D. and Gagnon, C. (2009) Conservation et écotourisme: Une lecture par la justice environnementale du cas des communautés voisines du Parc national de la Guadeloupe. *Études caribéennes*, April, 12.

Lowenhaupt Tsing, A. (2015) *The Mushroom at the End of the World: On the Possibility of Life in Capitalist Ruins*. Princeton, NJ: Princeton University Press.

Monjaret, A. and Pugeault, C. (eds) (2014) *Le sexe de l'enquête: Approches sociologiques et anthropologiques*. Lyon: ENS éditions.

Porter, B.A. and Schänzel, H.A. (eds) (2018) *Femininities in the Field: Tourism and Transdisciplinary Research*. Bristol: Channel View Publications.

Rauzduel, R. (1998) Ethnie, classes et contradictions culturelles en Guadeloupe. *SocioAnthropologie*, 4.

Shibutani, T. (1986) *Social Processes: An Introduction to Sociology*. Berkeley : University of California Press.

Vanderbeck, R.M. (2005) Masculinities and fieldwork: Widening the discussion. *Gender, Place & Culture* 12 (4), 387–402.

Wylie, A. and Sismondo, S. (2015) Standpoint theory in science. In J.D. Wright (ed.) *International Encyclopedia of the Social and Behavioral Sciences* (2nd edn) (pp. 324–330). New York: Elsevier.

9 Journeying into Yogaland: A Cautionary Tale of a White Guy's Perspectives on Yoga-related Fieldwork in Japan

Patrick McCartney

Globally, participation in yoga-inflected wellness tourism is predominantly female (~85%). It is even higher (95–99%) in Japan. Negotiating privileged access to the inner worlds of anyone is challenging; describing the embodied experiences of yoga consumers in Japan is even more difficult. How can one gain access to the private thoughts of people who are more interested in *doing* yoga as opposed to *thinking and talking* about it? This is exacerbated by being, like myself, white, male and poor at speaking Japanese. In essence, I am the very embodiment of a '*gaijin* (foreigner) smash' anthropologist, who constitutes the wilful or unwitting transgressions of Japan's social conventions.

While this book focuses on masculinities in the field, there are other issues that result in a dissonant coalescence of gender, such as language, culture, sexuality, religion, class, ethnicity, lifestyle and leisure. How can one deal with these issues discretely and collectively, especially the unforeseen problems that are more challenging to solve in the field? Through an auto-ethnographic perspective, I explore this case study of overcoming the problems of conducting research about yoga in Japan. This might prove useful to future researchers who find their positionality similarly problematic.

Introduction

How does one conduct anthropological research into yoga when one is male and global yogascapes can be exceptionally gendered spaces? Particularly when considering the added issue of being a white guy in Japan? *Gaijin smash* is a vernacular phrase referring to the ways in which foreigners (外人 *gaijin*) in Japan unwittingly or knowingly transgress local laws or social conventions. It is an apt descriptor of how I have in many ways, due to time restraints and lack of cultural knowledge, felt like I have clumsily collected data about yoga in Japan. While this is the question I attempt to answer in this chapter, I do not necessarily have sufficient answers. It is, rather, a cautionary tale of sorts, regaling my own shortcomings and subsequent on-the-fly attempts at solutions related to dealing with the unexpected exigencies of the field. In this chapter I first give a brief background of how I arrived in Japan, which is followed by the rest of the chapter explaining some of the issues I have faced and how I have worked proactively to deal with them.

Before Japan and the Gendered Reality of Global Yoga

Yogascapes in Japan is a two-year research project funded by the Japan Society for the Promotion of Science (JSPS), which is hosted by Kyoto University. It attempts to document the creation of demand and value-adding of global yoga-inflected lifestyles within the Japanese wellness and tourism markets. In plain speech, this translates into an attempt to document the biographies of Yoga and Sanskrit in order to understand how Japanese consumer society reconfigures many things Indian to suit its own aesthetic tastes, and how this relates to the global branding of yoga and the application of soft power strategies and cultural nationalisms.

I had not given much thought to yoga in Japan. To be honest, I had not given much thought to anything in Japan. For two decades at least, my obsession with South Asia had taken me to the subcontinent over 30 times. I have spent a total of over seven years living in India, first as a graduate student and then, after several more student visas, turning my time in India into a career. My experiences in India have shaped me and my understanding not just of yoga but also of masculinity. My understanding of the complexities of Japanese society and history is not at the same level as my intimate understanding of (north) Indian society and history. While I speak, read and write Hindi and Urdu as well as Sanskrit relatively fluently and, to a lesser extent, Marathi and Gujarati, breaking through the Japanese language barrier has been difficult. Even though I can speak, read and write basic, survival-level Japanese (日本語 *nihongo*), I have had an incredibly difficult time finding and/or retaining language exchanges, research assistants or translators to help with the task of interviewing people or becoming more familiar with Japan in general. Also, the way in which my research regarding Yoga and Sanskrit in India occurs is vastly different from the situation in Japan. For instance, my fieldwork in India has typically taken me to cloistered communities (*āśrama*-s) (McCartney, 2011, 2014, 2017a, 2018a) or small villages down dusty roads (McCartney, 2017b, 2017c, 2018b). People generally have time to hang out deeply in the Geertzian sense. In contrast, here in Japan it seems that many people are very busy. In addition, I have found the context of living in a big, modern city and using it as a site to conduct fieldwork quite distracting. Also, whereas I have personally found my interlocutors in India to be quite willing to talk, I have found the opposite to be the case in Japan. The reasons are multivariable. While every culture has this to a certain degree, the binary between an individual's true feelings (本音 *hon'ne*, 'true sound') and the coercive façade regulating expression of behaviour and opinions (建前 *tatemae*, 'built in front') results in frustrating attempts to elicit people's feelings. This is where the *gaijin smash* comes to the fore. I have probably been too impatient, seemingly wanting people to tell me everything about why they consume yoga in the brief moments we have together. But this is often the case because experience has told me that subsequent meetings will probably not occur, even if we become friends on various social media platforms. Also, globally speaking, consumers/practitioners of yoga are not necessarily particularly forthcoming about why they 'do yoga'. For many, yoga appears to be more of a verb than a noun, encompassing a sense of intransitivity or, rather, of *being-ness*, which conveys the sense of the Sanskrit term, *yogitva*. However, when such profound, numinous and cessative inner experiences result from the practice, it can be difficult to verbally articulate them (Sarbacker, 2008), especially to a stranger.

But this subjective, interior, epistemic relativism is an indelible part of Yogaland. This term appears around late 2004 (Google Trends, 2019). It is a euphemism often used disparagingly to critique the 'yoga industrial complex' (Broad, 2012) which is considered a colonial territory and place of entitlement (Russell, 2017). However, it is also a term that is used affectionately (McCartney, 2019a, 2019b, 2019c). And, even though Yogaland is generally considered a secular, hyper-commodified consumption-scape, the logic of its marketing, consumption and performance is inherently religious (Aechtner & Zambon, 2019; Foxen, 2017; Rinallo & Oliver, 2019). Prying into someone's 'yoga practice' is often considered either rude or distracting. It is considered to be 'sacred' and not necessarily something that ought to be studied, least of all by a white male perpetuating the perceptibly problematic colonial project of anthropology. Building on Coskuner-Balli and Ertimur (2016), who discuss legitimation strategies of hybrid forms of yoga, McCartney (2019b, 2020) discusses inner wellness tourism, internal and external yoga-related pilgrimage, imagined yoga consumption-scapes, and hybrid forms that combine, for example, Zen and Yoga through branding of 'Temple Yoga' (寺ヨガ *tera yoga*).

The topic of masculinity is a very interesting one, especially when we consider the layers involved in understanding its manifestation, performance and reception, not only within Japanese society but even more so within the glocalised domain of yoga. As Freeman (2017) shows, the overwhelming majority of famous Instagram yoga teachers in Japan are female. Understanding masculinity in Japanese society, particularly in relation to the ever-growing asexual identity for young Japanese men (草食系男子 *sōshokukei danshi*, 'herbivore boys'), whose emergence is more commonly referred to as a crisis in Japanese masculinity, is an interesting starting point (Saladin, 2015).

The growth in popularity of modern postural yoga (MPY) has its roots in the cultural hybrid of late 19th century colonial India, which saw bodybuilding 'athletes' become the promoters of a hyper-masculine yogic body that developed through the nascent Independence movement. This sought to counter the dominant narrative that Indian male bodies were perceived by the colonial powers as too effeminate (Newcombe, 2009). The political birth of modern yoga was thoroughly ensconced in promoting a 'muscular Hindu identity'. However, as Foxen (2017: 509) explains, 'This is to a large extent true insomuch as Yogi-figures are often portrayed in popular discourse in a rather paradoxical fashion as being interchangeably feminine and hyper-masculine'.

It was during the interwar period that yoga was exported into the West and became associated with a female physical culture, in contrast to the perceived masculine pursuit of body building. This perception endures. It is a major reason as to why participation rates across various geographic regions of the world are predominately female. Another way to perceive the popular conditioning related to commodified, modern, trans-global yoga – and how it is primarily marketed to and for women – is through a brief diachronic glance comparing the magazine covers of three prominent yoga-related magazines: (1) *Yoga Journal* (International edition, since 1975); (2) *Yoga Journal* (Japan, since 2008); and (3) *Yogini* (Japan, since 2004). At the time of writing, since the first volume of *Yoga Journal International* (May 1975), 362 covers have been published, of which 242 are freely available. Out of a total of 242 covers analysed, 174 feature only females, 41 feature only males, 16 feature both men and women and 11 feature some image other than a person (see Table 9.1). Out of 49 (of a total of 66) *Yoga Journal Japan* covers, 48 feature women, and out of 87 *Yogini Japan* covers analysed, 85 feature only women. The combined total is 307 out of 378 only featuring women on the cover. To view these magazine covers, see Yoga Journal (2019), Google Books (2019) and EI Publishing (2019). For comprehensive analysis of this topic, see Wittich and McCartney (2020).

Table 9.1 *Yoga Journal International* covers 1975–2019

	1970s	1980s	1990s	2000s	2010s	Totals
Male	9	16	11	3	2	41
Female	7	14	22	47	84	174
Mixed	2	5	7	1	1	16
Not specific	1	5	5	0	0	11

Currently valued at US$88 billion, the market size of yoga and Pilates studios is expected to increase at a compound annual growth rate (CAGR) of 11.7% between 2018 and 2025 to a staggering US$2.16 trillion (Bhandalkar *et al.*, 2019). The demographics for the average postural yoga class in the West vary between 70% and 85% women (Ibis World, 2015; Penman, 2008; Shift, 2016; Yoga Journal & Yoga Alliance, 2016; Yoga Journal, 2020).

Even though the participation of men in group yoga classes continues to increase, they are still predominately attended by women. When comparing this to yoga markets in East Asia, like those in China and Japan, women prefer yoga at a rate in excess of 90% (iResearch, 2018). It is worth noting that the *Yoga Journal*'s (2017) '日本のヨガマーケット調査/Japan Yoga Market Survey' shows a 69% predominance of women in yoga in Japan; however, this is likely attributed to the questionable recruiting methods used.

Anecdotal figures provided by the scores of yoga teachers I have personally spoken to in Japan puts the figure at least above 90%, if not higher. I have noticed that there is a higher participation rate among men at particular yoga festivals like Yogafest, Kyoto Yoga Day, Kobe Yoga Day and the International Day of Yoga. Shimizu (2014) places the male Yogafest figure at about 30% and I am inclined to agree with this figure. My observations also suggest that men are drawn to yoga as they age. I have noticed an almost 1:1 ratio between men and women at events that provide forms of 'yoga' based on singing devotional songs in the form of *bhakti* or *kīrtana*. Also, in the 50+ age category there is even more gender parity at devotional-style events that might not offer any MPY.

A common perception in Japan is that men who attend yoga classes possibly do so as a way to meet women or that they are gay. This is supported by blogs which ask whether men who practise yoga are gay (*homo*, ホモ) or transgender (*okama*, オカマ) (see Yogi, 2015). One fieldwork anecdote relates to an encounter I experienced in a yoga class in Osaka. The class of 20 people overwhelmingly consisted of 30–40 year old females, and included only one other male. After the class, he came over. 'Hi, my name is Hiro. I'm not gay! I just really like yoga.' My first question (to myself) was, 'Does he think I am gay?'

The founder of Tokyo Rainbow Yoga, Hiroshi Miyamoto, is a professional dancer who, after 20 years of living in Toronto, returned to Tokyo in 2013 (Rainbow Yoga Tokyo, 2019). His gay-friendly studio was a first in the country, even among only a handful of other yoga studios that offered yoga for men. Hiroshi feels that the general level of yoga advertising has neither increased nor declined over the past six years. However, he agrees that the prevailing opinion in Japan is that yoga is something women do. He also explained that his studio is small and that his customers use it more as a place to connect with other like-minded men.

Tarun Shekhar Jha is an Indian-born actor, model and yoga teacher who has lived in Tokyo since 2014. In 2019 he appeared in two popular lifestyle magazines. In August he appeared in *Tarzan*, which focuses on men's health (Kadogami, 2019). This particular article features photos of Tarun's muscled body in several poses. He explains how his muscle training/exercise regime is Hindu and suitable for yogis. As well, he claims it creates functional muscles and that the breathing practices (*prāṇāyāma*) reduce fatigue. It is an interesting example of yoga being promoted as an adjunct to muscle building that can aid recovery and help you to get bigger.

In contrast, he was next featured in *Yogini*, which is a lifestyle magazine heavily inflected by fetishised yoga accoutrements like mats, blocks and cushions (Yogini, 2019). I asked him about the perceptions of Japanese society towards men and yoga. He explained that being a male yoga teacher in Japan is challenging because of the enduring image that yoga is for females (see Kawasaki, 2013), and that in some of the places he teaches, he does not have access to any changing facilities.

In Japan

One way I understand the task, at least in part, of the anthropologist is to tune into the frequency of the culture that one ends up embedded in. After 22 months, it is fair to say that I do not feel like I have got close to knowing the bandwidth Japanese people are buzzing at. With every day that passes it becomes a little more familiar. Still, this causes me no end of anxiety when writing about yoga in Japan. However, the only way through this impasse is to keep talking to people, continue asking questions, keep an open mind, watch, look and listen and carry on reading about anything to do with Japanese society, history and culture. Many people might be initially interested in, as I frame it, 'starting a conversation', but I suspect that once they have had a chance to read some of my ideas on yoga, which may not necessarily align with theirs, they might reconsider and, ultimately, mute themselves.

One of the main hacks I have employed is to try not to speak to people. This might sound odd, coming from an anthropologist; however, it is not for lack of want or effort that I have resorted to this approach. Yoga people, particularly teachers, lead very busy lives. Many are mothers of young children trying to run their businesses. Therefore, pinning people down for even a short 10-minute chat, either face-to-face, via Skype or Zoom or by responding to emails, is challenging. When I do approach someone, either face-to-face or through social media, I normally begin by simply asking if they would like to chat about yoga with me, or not. While I might get a pleasant short introductory chat, it often proves difficult to build upon. Added to the mix is the extra issue of the prevalence of ambivalence, suspicion or hostility towards the academic study of yoga. This is not just an issue in Japan. There is a strong anti-intellectual sentiment that privileges knowledge claims of an epistemically subjective and, frequently, pseudo-scientific nature. As a general sceptic and critical realist, my 'red-pilling' of consumers of yoga has me often pigeon-holed as a heretic at best, and a 'white supremacist', 'neocolonialist', 'protector of white guilt' or 'silencer of women of colour' at worst. These are pejorative metonyms that have been hurled at me online from people outside Japan. Still, critical introspection is not necessarily the mainstay of the global yoga culture. Instead, yoga is an instrument, and perceptibly a good one at that, for re-enchanting disenchanted lives (Landy & Saler, 2009) and, particularly, as an individualised strategy for dealing with the demands of a post-Fordist, neoliberal labour market (Speier, 2019).

Despite this 'immersion' in Yogaland, my ability to sustain the level of cognitive dissonance and group thinking required within the cult-like, charismatic communities I moved through led to my scepticism. Now my relationship status with yoga is *complicated*. I have not taught a postural yoga class since 2015. I occasionally facilitate a workshop on chakras or something to that effect. The problem I have is that I cannot accept the influential node of Orientalising desires at the core of the global yoga industry and the Indian state, which mystifies yoga as some static, timeless, monolithic culture (Hebden, 2011). This is because of my critical thoughts on yoga and other intersecting issues, at which many people take offence, seemingly because of a reductive, binary, even fundamentalist-like approach to myth-making imagined communities (McCartney, 2017d, 2019a, 2019c). I have noticed that people often quickly settle into what feels like a sales pitch, in which they seem to favour spending our time trying to sell their brand of yoga. This can have a proselytising or missiological tenor (Lucia, 2018).

One example of an in-person meeting involves a rather prominent yoga teacher of Indian heritage, who was born and lives in Japan. Priyanka and I had communicated sporadically online for months. We finally met at a yoga event in Kobe. While it felt like I was in some way meeting an old friend, Priyanka was very interested in hearing my back story, my connection to yoga and how I position my work. This is a tricky moment for any anthropologist. In my case, quite often yoga producers and consumers want to assess quickly whether you are, or consider yourself to be, a fellow practitioner or not. And, if you are, then they want to know which tribe of yoga you belong to or identify with. This set of questions probably causes me more anxiety and guilt than any other. I run the risk of either coming across as a pompous know-it-all or someone with something to hide. Thus, my 'elevator pitch' is determined by who is asking for it. While they might seem like simple questions to be asked, 'Do you practise yoga?' or 'What sort of yoga do you do?', these questions, for me at least, are always loaded. However, if I am only asked the former question, then I can honestly reply in the positive, 'Yeah, sure, I do yoga', and breathe a sigh of relief if no further clarification is requested. I mean, yoga is very much a decentred, empty signifier, floating free in a transcultural current of the global consumption-scape (Fibiger, 2018; Okropiridze, 2018). When asked if I 'do yoga', my first response is to seek clarification about what, specifically, they mean by the term. I recently had someone say to me, 'you academics are so precise'. I took this as a compliment.

I do not really, as many modern postural yogis will claim, have 'an *āsana* practice'. While I do perform a lot of calisthenics and body weight exercises as part of my daily exercise regime, I struggle to identify that with an '*āsana* practice' (Foxen, 2017). If anything, I probably align more with the exercises explained in the 17th century wrestling compendium, the *Mallapurāṇa*(Sandhesara and Mehta, 1964; McCartney, *forthcoming*), which Tarun mentions that he includes, above, in his 'Hindu training'. However, when I am forced to think reflexively about yoga and what I might do, it ultimately has not much to do with physical exercise, meditation, devotional singing or attaining some lofty soteriological or theological goal. This is particularly so since the earliest attestation of the term 'Yoga' located in the *Ṛgveda* is found within a very martial context of contemporary early Ṛg Vedic tribal life in which *yudhmā viśaḥ* ('warring folks') engage in Yoga ('action') within the context of 'war' (RV 4.24.4) (Ananthanarayana & Lehman, 2016), which is the binary opposite to *kṣema* (periods of rest), for the sake of procuring prosperity and property (Palihawadana, 1968; Whitaker, 2011).

At some point Priyanka jumped in to proclaim that 'I think yoga is a lifestyle'. This is a common occurrence and refrain. It is one that many people in the industry use as part of a marketing strategy to generate lifestyle segmentation and convergence, particularly gender-based segments, that operate on a global scale (Damijanić, 2019; Myburgh *et al.*, 2019; Usunier & Stolz, 2014). Branding yoga in this way is an attempt to cultivate affect through the perception of value-adding via intangible goods. However, quite often, the 'way of life' invokes a holistic, totalising vision that is a fantasy outside of history. This becomes a particular site of contest between the 'precise' academic and the consumer/producer of a remarkably popular, albeit ahistorical, monolithic view of yoga. Yet, like so many other people, Priyanka's particular 'yoga lifestyle' is as arbitrary as any other in its appeals to efficacy, perceived cultural purity and authority, and mystery. And, seemingly, like any other producer of a yoga lifestyle, the assertion is that her brand (style of yoga, diet, ontology and epistemology) is the most effective. Still, the core of it, like so many others, represents the monolithic, static and, ironically, colonially constructed and Orientalist-imagined narrative about yoga's history, even if its aim is to decolonise yoga and attain some level of social justice through yoga (Antony, 2018; Godrej, 2016). In this situation, I have learnt to mute myself, temporarily suspend disbelief, and not get distracted by some argument about the historiographical details of yoga versus the popular imagination of yoga's history. The reason is that the production, consumption and expression of yoga, even within thoroughly 'secular' contexts, ought to be understood as operating through a religious logic (Jain,

2014; Lucia, 2018), which effects a ritual teleology of an existing religious system, misrecognising the interestedness to build revenue and brand loyalty (Foxen, 2017; Neumann, 2019).

One anecdote, which has frustratingly repeated itself many times, although not to the degree of the event I am about to explain, occurred in Nagoya. I was there to attend the Anthropology of Japan in Japan (AJJ) conference. My wife and I arrived the day before the conference to visit various yoga studios and to see the city.

Included on the itinerary was a hot yoga studio (ホットヨガ *hotto yoga*) which uses a heated stone (岩盤 *ganban*, stone) floor (Ganban Yoga, 2019). Ganban Yoga proposes that the style of heated rock used for the flooring emits negative ions and allows deeper penetration of the healing properties through the infra-red capabilities of the heating system, which apparently enables deeper detoxification. I was looking forward to visiting this studio, especially because it was a cold and windy day in early December. I have seen the look the receptionist gave me many times – the one that means 'you do not belong here'. Yoga studios are a particular cultural domain. I recall one occasion in New Delhi when a freshly minted (from a two-week yoga course in Rishikesh), early-20s Israeli woman declared that I could not be a yoga teacher because I did not look like one – even though I had taught yoga, Tai Chi, meditation and Sanskrit for close to two decades. Apparently, yoga teachers do not wear sports coats and collared shirts or have muscular frames. I was introduced to yoga as a child. Due to health reasons, a daily regime of stretching was prescribed by my physiotherapist. This, combined with a predisposition towards philosophy and historical linguistics converged through an intense study of Sanskrit, classical philology and archaeology, led me to where I am today.

Yoga studios, for many people, are a semi-sacred space, like the Pure Yoga studios in Hong Kong where busy and stressed 30–40 year olds, possibly 25% male and 60% Asian, were running out of elevators onto the 16th floor in suits and heels to quickly change into leggings for their blend of de-stress and fitness. Other patrons were sitting in the lounges glued to their phones or cursing at the receptionist and storming off over some issue. I briefly spoke to the manager, but once they realised I was not going to consume their product, their interest in our conversation abruptly evaporated. The inclination to talk, especially about the academic pursuit of *studying* yoga – as opposed to *doing* yoga – also often leads to very unfulfilling responses to questions. However, what quickly transpired in Nagoya involved the receptionist racing towards me while moving her arms into the X-position and repeatedly shouting *'dame'* (no). Forcibly pushed back through the automatic sliding door into the cold, breezy corridor with my boots now in my hands, I wondered how I could have done better. Ganban Yoga is like so many yoga studios around Japan. They only allow access to women. However, this was not clear to me, even though I made an effort to look for this information on the website.

Apart from the occasional death threat, attempts at doxing and threats of violence made towards my family, as well as attempts to sue me and take me to the Human Rights Commission for supposedly 'offending Hindus' or being slandered as a Mossad agent, most of which has occurred online or during fieldwork in India, this particular experience was possibly the most frustrating I have experienced in Japan. I just did not expect to experience a 'frontal push-out' sumo wrestling technique (押し出し *oshi-dashi*) during my fieldwork. This experience shook me up. I was confused, upset and embarrassed. I felt worse because I had *gaijin smashed* my way into evoking such a strong reaction in the receptionist. Godsoe (2016) compares criminal codes in different countries in relation to the legal consequences of Japanese subway pushers who forcibly cram patrons into train carriages, while Dean (2003) explains how Japan's management of society and settlement of disputes runs on the broad term, *giri* (義理), which conveys a sense of duty and social obligation. Although I did not feel physically threatened or consider pressing charges, I was a disturbance of the general harmony (和 *wa*) which necessitated the force used to remove me from the space (Dean, 2003).

In Japan, according to the *Yoga Journal* (2017a) survey, the reasons people start yoga are related to: (1) improving flexibility; (2) exercising for 'beauty and maintenance'; and (3) stress release and relaxation. I have personally witnessed a 95–100% female: male ratio.

A case in point is this 'Yoga, Travel & Wellness' opportunity for females to consume the 'Art of Femininity: 10 Day Women's Cultural Yoga Tour of Japan' (O'Dea, 2019). The promotional video frames the experience of yoga in Japan for tourists seeking wellness and yoga (O'Dea, 2017).

Most if not all the yoga teachers I have spoken to agree with this statistic. Yoga spaces in Japan are even more gendered than in other countries. I completely underestimated how this would impact on trying to conduct an ethnographic study. As the yoga teacher, Tom Wada, explains in an interview with *Metropolis* magazine:

> 'Now yoga is booming, especially among women, due to the growing number of actresses and models that are practicing and, hence, spreading the health benefit of it. But this creates yet another barrier', Tom explains. 'It's becoming more and more difficult for guys to enter a studio, because of the highly female-centred advertisements all around the city, giving the impression of yoga as a "girl thing"; whereas, in fact, for hundreds of years, it was solely a male practice'. Conceptually, Tom says the problem lies in the fact that men feel uncomfortable showing others their flexibility – especially to women. What is the solution to this modern-day gender problem in yoga? 'Many male celebrities regularly practice yoga and meditation. The only thing they need to do is "come out" to the public', Tom chuckles. 'So people will change the way they perceive it and think, "If he does it, I can do it, too!".' (Govoni, 2016, n.p.)

Not all yoga studios restrict access to women or have a tattoo policy. It is more likely that the boutique studios might be less concerned than the bigger commercial chains about tattoos. Businesses like the aforementioned Lava Yoga (2019), which has over 410 studios across the nation – most of which are women only – is similar to its smaller rival, CALDO, which has only 78 studios across Japan. However, CALDO explicitly states on the top of its website that it is a women-only space: 女性 (*josei*, women) 専用 (*sen'yō*, designated), which also offers ホット (*hotto*) ヨガ (yoga), ダイエット (*daietto*, diet), ボクシング (*bokushingu*, boxing) and ダンス (*dansu*, dancing) (CALDO Hot Yoga, 2019). If one looks at the client base, marketing strategy and demographic that these businesses are seeking to value-add 'yoga lifestyles' to (Sola Studio, 2019), it becomes quite clear that the demographic is one that is similar to what we see in many yoga-related markets around the world. A poignant example is one advertisement for Lava Yoga (@knto_rin5 (2018), which featured on many billboards and subway platforms that presents an unexpected example of trans-speciesism. Of the many people I asked, none were able to provide meaningful explanation as to why one of the women wears a mule head costume. The main text, reads なぜラバ (*nanze raba?*; Why Lava?). This is a pun. ラバ (*raba*, mule) is spelled the same way that lava (ラバ, *raba*) is pronounced and transliterated. It is a curious advertising campaign. Will people turn into asses or mules if they attend Lava Yoga?

Another issue – which really set back my initial project plan – relates to the cost of participating in yoga lifestyles in Japan. Yoga is not cheap, generally speaking, no matter where it is practised. For all the assertions that 'yoga unites', there are many countering assertions that show how yoga can also lead to separation, and regardless of the platitudes asserting that Yoga brings people together, there are many instances where it creates social boundaries through being a form of high-brow culture that creates class structured opposition (McCartney, 2016, *forthcoming*; Reeves 2019). The price of consuming yoga classes in Japan is quite high, with the average price per class being around jpy2500 (US$24). The cheapest class I have seen is jpy800 (US$8) for a 'chair yoga' class at the local community centre (Banno, 2018). Some free events are also offered for promotional purposes, such as an event held in Kyoto as a trial run to see how yoga tourism could grow (Kyoto City Official Travel Guide, 2019). Some yoga classes can cost jpy4000 (US$35) or even more. With a modest research budget that failed to cover a significant amount of my proposal, the work evolved to incorporate using social media and the publishing of yoga-related promotional videos on YouTube and Vimeo as a steady source for data, instead of trying to attend more yoga classes. Also, having arrived in Kyoto, I was then told by the administration managing my research fund that I could not submit receipts for yoga classes because this was deemed not to be part of my research. This took over a month to solve.

I initially intended to visit as many studios in the Kansai area as possible, hoping to find a suitable location to embed myself and carry out a focused ethnographic study. However, this did not materialise in the way I had hoped. Many yoga studios in Kyoto, for instance, were less than inclined to have me attend with the primary intention of conducting research. It was considered to be an intrusion that the customers should not have to bear. And, as I have learnt, anything less than an enthusiastic 'yes' is really a polite 'no' (see Porter, 2018, for an interesting vignette relating to this). Spirit Yoga Studio in Osaka welcomed me, but over time I grew fatigued with the 90-minute commute each way. Also, I found that people were quite reluctant to speak to me or they were just too pressed for time. Not just at Spirit Yoga, but on many occasions, it became routine that I was just attending a yoga class for my own benefit.

An example of exploring the popular media representations of yoga can be found through NHK (the national broadcaster), such as this clip which discusses the International Day of Yoga (Embassy of India in Japan, 2016). Another example is the ヨガ (yoga) 婚活 (*kon katsu*, marriage) セミナー (*seminā*, seminar) & お見合い (*O miai*, matchmaking) event (Ringo no Hana, 2019), which actively encourages men to participate – however, at an inflated price. Men are required to pay jpy3500 (US$31), whereas women are only required to pay jpy2500 (US$22). The more pertinent problem is that one must be single and actively looking for a life partner to marry in order to attend this event. Even though I once applied to participate – before I realised this rule – the event was actually cancelled because not enough men applied. I certainly considered lying to attend subsequent events. I also introduced the opportunity to some of my male and female friends who are single, citing that it could be good for us both if they attended. None agreed, however. It seems that Japanese men are not as interested in yoga-pants-wearing yoginis as their Western counterparts seem to be. Also, the rarer commodity in Japan's Yogaland is men. It would seem that one way to encourage more men to participate would be by charging less. Regardless, I tried to contact the organiser through Line, Facebook, email, etc. in an attempt to have a quick chat about the basic details of the event. These requests for anonymous metadata (e.g. how it started, when, who typically comes, their ages, professions, interest in yoga) were met with silence. My work-around involved reading through the archives of the website. I found a report for the first 'yoga matchmaking seminar' in which five men and four women participated. From this group two couples formed and one pair has since had a child. The use of yoga, we are told, is to help people relax and become more open. It seems that some people share this sentiment, since we are able to see on the comments section of the website, 'It was good that I could talk while relaxing after having done yoga' (Ringo no Hana, 2019).

Any time I come across something relevant, I write to the organiser of the event or business. I subscribe and follow all their social media channels. I begin 'liking' their pages and some of their posts. I will then ask the individual, usually through Messenger or something similar, if they would be interested in having a short chat about yoga and wellness. I explain what I am doing and why. Most of the time I simply want to know how big their studio is, how long it has been open for, what the gender, age, ethnic and class demographics are, if they have been or intend to go/return to India, and what their definition of yoga is.

I believe that anthropologists ought to pay attention to social psychology and the processes underlying persuasion, particularly from a discursive interactional infrastructure perspective of sales and marketing – and at least the techniques employed to build rapport. Relying on discursive psychology and conversational analysis, Humă *et al.* (2018) identify two sets of communicative practices that boil down to promoting *alignment* and hampering *resistance*. Industry reports show that when talking to people in a sales context, factors that influence 'buyer purchase decision' include the value delivery, potential benefit of collaborating, education on new ideas and opportunities, and deepening understanding of needs within an industry (Schultz *et al.*, 2019). Social scientists are *selling* intangible goods. We seek cooperation. Some of the things I have done to entice participation include offering to make a short three-minute promotional video, write a 1000-word review of an event or give a free talk about some aspect of yoga. These strategies have had varying levels of success. There are particular narratives and ontologies within Yogaland that are in many ways incommensurable with a scholarly epistemology.

I am constantly trying to find innovative ways to make my research appealing. Despite this, perhaps only 1% of yoga people respond and of that, perhaps 0.5% maintain some consistent contact. This contact, however, is at risk of being lost if I say something that runs counter to their more rigid, doctrinal, Orientalist-imagined narratives of yoga's history. I feel ethically bound to assert that a particular point is wrong, even if it might offend the other person. Should I support ignorance and stay quiet in order not to upset my interlocutor? I always find it difficult not to point out, for example, that a particular detail about yoga's history is not typically accepted to be true by Sanskrit scholars who read the texts in the original, and do not just rely on a poorly researched yoga teacher-training manual or blog post to ascertain the veracity of some truth claim.

Perhaps I was too naïve to think that if I showed up, took a class or two, liked some posts on a social media page, smiled and generally tried to just fit in, then I would be accepted. I feel that I was quite unprepared to conduct research on yoga in Japan. While my ethnicity, gender, and competency in Japanese have certainly impacted the outcomes, the governing practice and logics of this field have also contributed in different ways. Had I known all this in advance, then I would definitely have approached it differently. I might even have chosen a completely different topic and location.

In order to get around this disappointing rate of rapport and participation, I had to rethink my research methods and aims. The way forward is to talk to as many people as possible. The other path chosen has been through a multimodal analysis of visual assemblages, media content, discourse and sentiment analysis, as well as transcribing embodied action through watching promotional videos, from which I then build up an analytical template (Luff & Heath, 2016). Subscribing to social media pages keeps me updated on promotions and events. I have also set up online surveys and sent them out in Japanese and English. And I have manually or automatically *scraped* publicly available information from the internet. Two examples including this scraped information relate to visually representing the spatial distribution of yoga studios across Japan (McCartney, 2019d) and offering an anonymous way to engage through a 'yoga survey' (McCartney, 2019e). What this map and survey tell us is that yoga in Japan, as in many other places around the world, is a mostly urban phenomenon consumed by women. The reason for offering the survey online in both Japanese and English is to allow people to opt in to providing information in a less threatening way and avoid the issues and confusion that often arise from disclosing information in person.

During my time in Japan I have moved into more macro-level analysis of events and networks; this includes a keener focus on big data. However, this emerging field has its own ethical issues. Would I have become more sociological and macro-level focused had I not moved to Japan? The main issue of not understanding the gendered space of yoga studios in Japan is certainly compounded by not necessarily feeling welcome in many spaces. Whether this is because of my gender and ethnicity is, at this point, irrelevant. However, my poor language competency, the relatively small size of yoga studios which does not allow people to 'hang out', and the fact that people are incredibly busy and also sceptical of academic scholars of yoga has certainly pushed me to think about how to say something meaningful in ways beyond the micro-level constraints outlined above. I have become in some ways less interested or bothered by what people might say to me in person or online. Teetering between scepticism and cynicism, I find the types of response bias seemingly impossible to filter out, particularly when it comes to social desirability bias. People are unreliable and might not even be able, or want, to articulate why they do what they do. Instead, they say what they think you might want to hear, even in anonymous surveys (Jann et al., 2019); however, Nederhof (1985) has developed some coping strategies. I have found comfort in focusing more on industry reports in relation to the consumptive practices.

Concluding Remarks

I feel I have only just scratched the surface of how being a male in the overwhelmingly gendered space of yoga studios, particularly in Japan, presents many issues to address. I do not necessarily consider that I have found the best ways to deal with the issues discussed. Hopefully this discussion sheds light on some of the possible ways one can work around the issues relating to masculinities in the field. There are, as I continue to learn, many ways to be an anthropologist and to do anthropology. Prior to arriving in Japan my plan was essentially to travel to different cities like a wandering nomad and spend as much time as possible in a few different studios. The financial cost of this peripatetic choice quickly eroded any attempt to pursue it. Also, the difficulty in finding willing and tolerant studios in which to conduct focused ethnographies has resulted in my changing the way I approach the fieldwork. This has as much to do with my fellowship's funding regulations as it does with the small size of yoga studios, which makes it hard for people to mill about before or after classes. As well, the indelible fact that yoga studios are typically, for better or worse, urban sanctuaries for women seems an intractable obstacle for someone like myself. This is for good reason. Regrettably, like in many places around the world, gendered safe spaces are necessary in Japan. The female-only train carriages are meant to enable women and girls to travel without being assaulted or, more specifically, groped in a crowded train carriage. The concept of 痴漢 (*chikan*, sexual harassment) is a social issue for any woman aged between 10 and 50 years (Horii & Burgess, 2011; Lee, 2017), even though the Japanese media promotes the idea that it is the 'safest country in the world' (Arudou, 2017). Based on Horii and Burgess (2011), we can infer that the yoga studio in Japan becomes not only a space to potentially avoid *chikan* or to have respite from the anxiety of, and potential for, being physically groped, but also as a locus for the symbolic rejection of a particular type of masculinity. While not focusing on the possibility of inappropriate touching and other coercive or violent acts of a predatory kind that can occur within the field of yoga, in any time or place, the way in which this has all compounded being a white guy trying to *gaijin smash* my way into conducting anthropologically informed research about yogascapes in Japan has led me on some unexpected paths. Hopefully, these musings will help the reader to think outside the box and to deal practically and creatively with similar situations that will, inevitably, arise.

References

@knto_rin5 (2018) #LAVA. *Twitter*. See https://twitter.com/knto_rin5/status/1035747425846145029.

Aechtner, T. and Zambon, O. (2019) Convergent antievolutionism and the Hare Krishnas. *Theology and Science* 17 (3), 292–296. doi:10.1080/14746700.2019.1632516

Antony, M.G. (2018) That's a stretch: Reconstructing, rearticulating, and commodifying yoga. *Frontiers in Communication* 3 (October), 1–12. doi:10.3389/fcomm.2018.00047

Arudou, D. (2017) Media marginalization and vilification of minorities in Japan. In J. Kingston (ed.) *Press Freedom in Contemporary Japan* (pp. 213–228). New York: Routledge.

Askegaard, S. and Eckhardt, G.M. (2012) Glocal yoga: Re-appropriation in the Indian consumptionscape. *Marketing Theory* 12 (1), 45–60. doi:10.1177/1470593111424180

Banno, M. (2018) *NHKテレビ　Masuのヨガスクワット!*. See https://www.youtube.com/watch?v=KZPwQkJ0wR0.

Bhandalkar, S., Das, D. and Kadam, A. (2019) *Pilates and Yoga Studios Market by Activity (Yoga Classes, Pilates Classes, Pilates and Yoga Accreditation Training, and Merchandise Sales): Global Opportunity Analysis and Industry Forecast, 2018–2025*. See https://www.alliedmarketresearch.com/pilates-and-yoga-studios-market.

Broad, W.J. (2012) *The Science of Yoga: The Risks and Rewards*. New York: Simon & Schuster.

CALDO Hot Yoga (2019) *Women-only Hot Yoga*. See https://www.hotyoga-caldo.com/shijoomiya/.

Coskuner-Balli, G. and Ertimur, B. (2016) Legitimation of hybrid cultural products: The case of American Yoga. *Marketing Theory* 17 (2), 127–147. doi:10.1177/1470593116659786

Damijanić, A.T. (2019) Wellness and healthy lifestyle in tourism settings. *Tourism Review* 74 (4), 978–989. doi:10.1108/TR-02-2019-0046

Dean, M. (2003) *Japanese Legal System: Text, Cases and Materials*. London: Cavendish.

EI Publishing (2019) *Yogini Magazine*. See https://www.ei-publishing.co.jp/magazines/regular/yogini/page/2/.

Embassy of India in Japan (2016) International Day of Yoga 2016 in Japan. *NHK News Clip*, 24 June. See https://www.youtube.com/watch?v=FlYsG8Wbkxo.

Fibiger, Marianne Qvortrup (2018) "Śrī Mātā Amtānandamayī Devī—The Global Worship of an Indian Female Guru." In Jørn Borup and Marianne Qvortrup Fibiger (eds) *Eastspirit: Transnational Spirituality and Religious Circulation in East and West* (pp. 80–99). Leiden: Brill.

Foxen, A.P. (2017) Yogi calisthenics: What the 'non-yoga' yogic practice of Paramahansa Yogananda can tell us about religion. *Journal of the American Academy of Religion* 85 (2), 494–526. doi:10.1093/jaarel/lfw077

Freeman, E. (2017) *19 Instagram Yogis in Japan*. See https://medium.com/@ellenlouise-freeman/instagram-yoga-teacher-photography-models-japan-6e189e80de9c.

Ganban Yoga (2019) *Bedrock Hot Yoga*. See https://relaxation-sola.co.jp/yoga/.

Godrej, F. (2016) The neoliberal yogi and the politics of yoga. *Political Theory* 45 (6), 1–29. doi:10.1177/0090591716643604

Godsoe, H. (2016) The law of Japanese subway pushers. See https://jurisjapan.com/tag/assault/.

Google Books (2019) *Yoga Journal: 1970–2000*. See https://books.google.com.hk/books/serial/ISSN:01910965?rview=1.

Google Trends (2019) *Yogaland*. See https://trends.google.com/trends/explore?date=all&q=yogaland.

Govoni, N. (2016) The yoga man. *Metropolis*, 6 June. See https://metropolisjapan.com/the-yoga-man/.

Hebden, K. (2011) *Dalit Theology and Christian Anarchism*. Farnham: Ashgate.

Horii, M. and Burgess, A. (2011) Constructing sexual risk: 'Chikan', collapsing male authority and the emergence of women-only train carriages in Japan. *Health, Risk and Society* 14 (1), 41–55. doi:10.1080/13698575.2011.641523

Humă, B., Stokoe, E. and Sikveland, R.O. (2018) Persuasive conduct: Alignment and resistance in prospecting "Cold" calls. *Journal of Language and Social Psychology* 38 (1): 33–60. doi:10.1177/0261927X18783474

Ibis World (2015) *Pilates and Yoga Studios in Australia: Market Research Report*. See http://www.ibisworld.com.au/industry/pilates-and-yoga-studios.html.

iResearch (2018) *2018 China's Yoga Industry Research Report*. See http://www.iresearch-china.com/content/details8_49385.html.

Jain, A.R. (2014) *Selling Yoga: From Counterculture to Pop Culture*. Oxford: Oxford University Press.

Jann, B., Krumpal, I. and Wolter, F. (2019) Editorial: Social desirability bias in surveys – collecting and analyzing sensitive data. *Methods, Data, Analyses* 13 (1), 3–6.

Kadogami, N. (2019) He has experience in karate, muay thai, wrestling and body building. Yoga teacher's muscles. *Muscle Picture Book, Vol. 6* [空手やムエタイ、レスリング、ボディビルディングの経験もあり。ヨガ講師の筋肉｜筋肉図鑑 vol. 6]. See https://tarzanweb.jp/post-195483?fbclid=IwAR2h2GvZLR3JJusVPskWLzgQmM7pTMyuyH9xwPhJIH_WosKRmDbaEocA1BQ.

Kawasaki, K. (2013) 石ちゃん、カルド川崎のホットヨガに挑戦！ *Ishi-chan Challenges Kardo Kawasaki's to Hot Yoga!* See https://www.youtube.com/watch?v=jCgiQlG0OHU.

Kyoto City Official Travel Guide (2019) *Yoga Experience at the Old Mitsui Family Shimogamo Villa*. See https://prev.kyoto.travel/ms/latest_news2/133.

Landy, J. and Saler, M. (2009) Introduction: The varieties of modern enchantment. In J. Landy and M. Saler (eds) *The Re-Enchantment of the World: Secular Magic in a Rational Age* (pp. 1–14). Stanford, CA: Stanford University Press.

Lava Yoga (2019) *Lava Hot Yoga Studio*. See https://yoga-lava.com/.

Lee, A. (2017) Gender, everyday mobility, and mass transit in urban Asia. *Mobility in History* 8 (1), 85–94. doi:10.3167/mih.2017.080110

Lucia, A. (2018) Saving yogis: Spiritual nationalism and the proselytizing missions of global yoga. In B.E. Brown and B.S.A. Yeoh (eds) *Asian Migrants and Religious Experience: From Missionary Journeys to Labor Mobility* (pp. 35–70). Amsterdam: Amsterdam University Press.

Luff, P. and Heath, C. (2016) Transcribing embodied action. In D. Tannen, H.E. Hamilton and D. Schiffrin (eds) *The Handbook of Discourse Analysis, Vol. 1* (2nd edn) (pp. 367–390). Chichester: Wiley Blackwell.

McCartney, P. (2011) Spoken Sanskrit in a Gujarat ashram. *JOSA* 43 (Australasian Sanskrit Conference Special Issue), 61–82.

McCartney, P. (2014) *The Sanitising Power of Spoken Sanskrit*. See http://himalmag.com/sanitising-power-spoken-sanskrit/.

McCartney, P. (2016) Utopian symmetries: Reflections on future worlds and transglobal yoga. *Journal of the International Society for the Interdisciplinary Study of Symmetry* 1, 86–89.

McCartney, P. (2017a) Suggesting *Śāntarasa* in Shanti Mandir's *Satsaṅga*: Ritual, performativity and ethnography in Yogaland. *Ethnologia Actualis* 17 (2), 81–122. doi:10.2478/eas-2018-0005

McCartney, P. (2017b) Jhirī: A 'Sanskrit-speaking' village in Madhya Pradesh. *Journal of South Asian Languages and Linguistics* 4 (2), 167–209. doi:10.1515/jsall-2017-0007

McCartney, P. (2017c) Speaking of the little traditions: Agency and imposition in 'Sanskrit-speaking' villages in north India. In L. den Boer and D. Cuneo (eds) *Puṣpikā: Tracing Ancient India Through Text and Traditions* (pp. 62–88). Philadelphia, PA: Oxbow Books.

McCartney, P. (2017d) Politics beyond the yoga mat: Yoga fundamentalism and the 'Vedic way of life'. *Global Ethnographic* 4, 1–18.

McCartney, P. (2018a) Śāntamūrti: The legitimate disposition(s) of the 'Temple of Peace' social network. *Bulletin of the Nanzan Institute for Religion and Culture* 8, 65–104. doi:10.17605/OSF.IO/4MP3K

McCartney, P. (2018b) Notes from the field: Reflections on the Imagining Sanskrit Land Project. *Global Ethnographic* 4, 2–9.

McCartney, P. (2019a) Spiritual bypass and entanglement in Yogaland (योगस्तान): How neoliberalism, soft Hindutva and banal nationalism facilitate yoga fundamentalism. *Politics and Religion Journal* 13 (1), 137–175. doi:10.17605/OSF.IO/MFB5N

McCartney, P. (2019b) Yoga-scapes, embodiment and imagined spiritual tourism. In C. Palmer and H. Andrews (eds) *Tourism and Embodiment* (pp. 86–106). Abingdon: Routledge.

McCartney, P. (2019c) Stretching in the Yogaland's shade: Unlikely alliances, strategic syncretism and de-post-colonizing Yogatopia(s). *Asian Ethnology* 78 (2).

McCartney, P. (2019d) *Yoga Studio Maps*. See https://www.yogascapesinjapan.com/yoga-maps.html.

McCartney, P. (2019e) *Yoga Survey*. See https://www.yogascapesinjapan.com/yoga-survey.html.

McCartney, P. (2020) The X + Y + Zen of "Temple Yoga" in Japan: Heretically-sealed cultural hybridity. *Journal of Dharma Studies*: 1-16. https://doi.org/10.1007/s42240-020-00069-9.

McCartney, P. (*forthcoming*) Poles apart? Haṭhayoga, Mallakhamb, and the Ancient Origins of Pole Yoga. In M. Singleton and D. Bevilacqua (eds) *Yoga and the Physical Practices of South Asia*. Cambridge: Open Book Publishers.

McCartney, P. (*forthcoming*) The abuse of yoga and Proposition YSB. In C.P. Miller and J. Hargreaves (eds) *The Luminescent: Abuse in Yoga and Beyond Special Issue*.

Myburgh, E., Kruger, M. and Saayman, M. (2019) When sport becomes a way of life: A lifestyle market segmentation approach. *Managing Sport and Leisure* 24 (1–3), 97–118.

Nederhof, A.J. (1985) Methods of coping with social desirability bias: A review. *European Journal of Social Psychology* 15 (3), 263–280. doi:10.1002/ejsp.2420150303

Neumann, D.J. (2019) Development of body, mind, and soul: Paramahansa Yogananda's marketing of yoga-based religion. *Religion and American Culture: A Journal of Interpretation* 29 (1), 65–101. doi:10.1017/rac.2018.4

Newcombe, S. (2009) The development of modern yoga: A survey of the field. *Religion Compass* 3 (6), 986–1002. doi:10.1111/j.1749-8171.2009.00171.x

O'Dea, B. (2017) *Art of Femininity Women's Cultural Yoga Tour Japan*. See https://www.youtube.com/watch?v=y3o_jHQLwQQ&feature=youtu.be.

O'Dea, B. (2019) *Art of Femininity*. See https://belindaodea.com/womens-yoga-holiday-japan/.

Okropiridze, D. (2018) Restricted access 'East' and 'West' in the kaleidoscope of transculturality: The discursive production of the Kualinī as a new ontological object within and beyond Orientalist dichotomies. In J. Borup and M.Q. Fibiger (eds) *Eastspirit: Transnational Spirituality and Religious Circulation in East and West* (pp. 120–145). Leiden: Brill.

Penman, S.J. (2008) Yoga in Australia: Results of a national survey. MA dissertation, RMIT University Melbourne.

Porter, B. (2018) The yes-no-ma'amsir blur. In P. Mura and C. Khoo-Lattimore (eds) *Asian Qualitative Research in Tourism Ontologies, Epistemologies, Methodologies, and Methods* (pp. 305–306). Singapore: Springer.

Rainbow Yoga Tokyo (2019) *Rainbow Yoga Tokyo*. See https://rainbowyogatokyo.com/?fbclid=IwAR2rr0pP-YOYJKAz2eG9pu8TGiNu3MtujKH7SSnjvcsnvSBN4mhpmL3-eI0.

Reeves, A. (2019) How class identities shape highbrow consumption: A cross-national analysis of 30 European countries and regions. *Poetics* 76 (October), 1–43, doi:10.1016/j.poetic.2019.04.002

Rinallo, D. and Oliver, M.A. (2019) The marketing and consumption of spirituality and religion. *Journal of Management, Spirituality and Religion* 16 (1), 1–5. doi:10.1080/14766086.2019.1555885

Ringo no Hana (2019) *Yoga 婚活セミナー&お見合い第2弾*. See https://www.ringonohana-enmusubi.jp/.

Russell, J. (2017) *Cultural Appropriation Part 1: India*. See https://www.jamesrussellyoga.co.uk/blog-james-russell_files/yoga-cultural-appropriation.html.

Saladin, R. (2015) Between gyaru-o and sōshokukei danshi: Body discourses in lifestyle magazines for young Japanese men. *Contemporary Japan* 27 (1), 53–70. doi:10.1515/cj-2015-0004

Sandesara, B.J. and Mehta, R.N. (1964) *Mallapurana*. Pune: BORI.

Sarbacker, S.R. (2008) The numinous and cessative in modern yoga. In M. Singleton and J. Byrne (eds) *Yoga in the Modern World: Contemporary Perspectives* (pp. 161–183). London: Routledge.

Schultz, M., Murray, J. and Smith, G. (2019) *Sales Prospecting Myths Debunked*. Boston, MA: Rain Group.

Shift (2016) *Infographic: The Growing Yoga Industry – Our Insights and Conclusions*. See https://www.shift.is/2016/04/infographic-growing-yoga-industry-insights-conclusions/.

Shimizu, K. (2014) The yoga business: Spreading beyond the boundaries of age & sex. *Japan Economic Forum* 198 (8), 30–33.

Sola Studio (2019) Yoga to live simply – yoga is a way of life. *Sola Blog*, 11 May. See https://relaxation-sola.co.jp/blog/post-2145/.

Speier, A. (2019) Yoga as an embodied journey towards flexibility, openness and balance. In C. Palmer and H. Andrews (eds) *Tourism and Embodiment* (pp. 71–85). New York: Routledge.

Usunier, J.C. and Stolz, J. (eds) (2014) *Religions as Brands: New Perspectives on the Marketization of Religion and Spirituality*. Farnham: Ashgate.

Whitaker, J. (2011) *Strong Arms and Drinking Strength: Masculinity, Violence, and the Body in Ancient India*. New York: Oxford University Press.

Wittich, A. and McCartney, P. (2020) The changing face of the yoga industry, its dharmic roots, and its message to women: An analysis of yoga journal magazine covers, 1975–2019. *International Journal of Dharma Studies* 3 (1), 1–14. doi:10.1007/s42240-020-00071-1

Yoga Journal (2017) 日本のヨガマーケット調査/*Japan Yoga Market Survey*. Tokyo: Seven and i Holdings.

Yoga Journal (2019) *Yoga Journal Library*. See https://www.yogajournallibrary.com/browse/2000/.

Yoga Journal (2020) *Media Kit 2020*. See https://www.yogajournal.com/page/advertise.

Yoga Journal & Yoga Alliance (2016) *2016 Yoga in America Study*. See https://www.yogajournal.com/page/yogainamericastudy.

Yogi, M. (2015) There are a lot of guys doing yoga, that is yogis are gay (homo or okama?) [ヨガやっている男、つまりヨギってゲイ (ホモ? オカマ?) 多いよね]. See https://ameblo.jp/musashino-yoggy/entry-12020598039.html.

Yogini (2019) Anatomy of the heart [心の解剖学; *kokoro no kaibō-gaku*]. *Yogini* 71. See https://www.amazon.co.jp/YOGINI-ヨギーニ-VOL-71-2019年9月号-Yogini編集部/dp/B07T5R65M1/ref=sr_1_1?camp=247&creative=1211&keywords=Yogini&linkCode=ur2&qid=1567676652&s=books&sr=1-1.

10 A *Mzungu* in Kenya: Dissonant Masculinity and Ethnographic Field Research in Sub-Saharan Africa

Gary Lacey

Introduction

Ethnographic field research involves the immersion of the researcher into the community or communities being studied, thus exposing the researcher's differing cultural traits, including masculinity. Ethnographic methodology was developed by anthropologists who would often spend many years immersed in a community, attempting to develop an insider's perspective on the culture (Tedlock, 1991). Such efforts are generally recognised as being imperfect since it seems to be virtually impossible for an immersed outsider to shake off his or her culture entirely. White settlers who adopted African ways were said to have *gone native*, yet they did not completely fit the African mould, even when accepted as part of an African tribe (see Ricciardi, 1981, quoted in Huxley, 1997). Anthropologists who remain permanently in their study communities are also said to *go native*. As with the colonist, the anthropologist who goes native typically retains elements of foreignness, never entirely losing the alien identity (Tedlock, 1991).

Tourism ethnographers seldom have the luxury of spending many years immersed in their research cohort's culture. Ethnographic methodology is often applied for just a few months or even a few weeks. In such a setting, shedding one's own habitus is extremely problematic. Habitus refers to elements of one's culture and upbringing that tend to shape the way one thinks and behaves. Such things as religion, morality, education, even physicality (racial markers, hairstyle, body adornment, etc.) can be regarded as elements of habitus. Consequently, it has become de rigueur for ethnographic studies to be accompanied by a statement about the author's habitus. Not only does such a statement allow the reader to judge how cultural factors have affected the study's conclusions, but its construction also forces the researcher to reflect on the interaction between his or her habitus and the culture of the study community.

Gender, and the way we display our gender identity, can influence both the way we interpret our surroundings and the degree to which we are accepted by those with whom we interact. But gender expression, in this case masculinity, is not an independent factor. It is highly interactive and cannot be meaningfully explored in total isolation from other cultural elements. Accordingly, through experiences in tourism and social science research in Central and Western Kenya, this chapter explores masculinity in African field research as an element acting upon, and being acted upon by, numerous aspects of the researcher's and the host community's cultures. These will be reflected upon through my PhD study into philanthropic tourism in the Nyeri district in Central Kenya in 2008/2009, and through two studies (one into HIV education and the other into county policing and town planning) conducted in Kisii, Western Kenya, in 2015 and 2018, respectively (see Figure 10.1). Experiences related to tourism research in Botswana are also drawn upon.

My habitus

Much of my own research has been conducted in sub-Saharan Africa, particularly Kenya and Botswana. Kenya has been the focus of my field

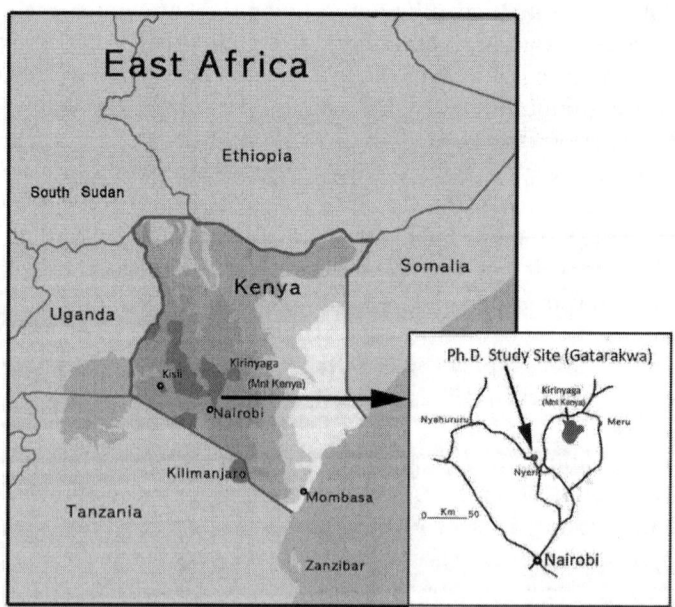

Figure 10.1 Map of Kenya showing PhD study site and Kisii
Source: Author.

experiences since my research in Botswana has, to date, been conducted collaboratively with a female colleague from the University of Botswana. It is fitting, therefore, to outline, briefly, some key elements of my own habitus pertaining to both social science and tourism research in Kenya. While I am an Australian citizen and resident, I was born in the coastal Kenyan city of Mombasa during the Mau Mau uprising in the 1950s. Although I have mixed heritage, being of Indian and Irish descent on my father's side and Scottish and German-Jewish on my mother's, I have the appearance of a white, ethnic European (see Figure 10.2). Consequently, I am (to use the Kiswahili word) a *Mzungu*, although I am sometimes referred to by Kenyans as an Indian (*Mhindi*) and more rarely as a Swahili (*Mswahili*). My early childhood was spent in Kenya but the vast bulk of my life, and my schooling, have been in Australia. I was therefore raised on the philosophy of the 1960s' and 1970s' second wave of feminism. In Australia, I was raised with a mixture of cultural elements, including Indian food and a mix of Kiswahili, Hindi and English languages spoken at home. Stories of my father's life in Rangoon and Darjeeling were intertwined with recollections of Kenya and Uganda (where my brother was born), as well as tales about my mother's Munro clan identities and her father's Jewish heritage. Above all, I maintained a strong Kenyan and East African identity. In my adult years I became active in the East and Central African diaspora in Melbourne. Today, I have a house and family in Kenya, but during most of the fieldwork discussed in this chapter I was single and childless. Since sexuality is an element that can affect

Figure 10.2 At a cattle post in Botswana; being comfortable around cattle is an important cultural alignment with many sub-Saharan African communities
Source: Author.

masculinity perceptions, it is worth noting that I am a heterosexual, cis male (born and identifying as male, see Couric, 2017). My mixed racial heritage and birth in Kenya have resulted in my ethnic and national identities also being mixed. I regard myself as a Kenyan-Australian, and as partly Indian, Celtic and Jewish. It is also worth noting, in the context of African research, that I am an atheist, with no spiritual beliefs.

Masculinity in Australia and Kenya

Masculinity (and its counterpart, femininity) are viewed as social co-constructs, separate from, but often related to, physicality (Connell, 2005). In other words, gender is related to masculinity/femininity, but one is a biological condition and the other is social (Couric, 2017). Our gender biology has not changed significantly, but societies have evolved. Hence, it can be argued that there is a growing separation between being male and being masculine. This has led many to observe that there is a crisis of masculinity in Western countries (Beynon, 2002; Forth, 2008). Indeed, gender identification has become a complex issue, with Facebook allowing users to choose either a binary gender (male/female) or a customised gender identity, which can include cis male or cis female, transgendered, intersex, etc., as well as the type of pronoun one wishes used (he/she/they, etc.) (Couric, 2017).

For the non-hegemonic male, the way masculinity is displayed might prove problematic (Vanderbeck, 2005). In a society that defines masculinity and gender in restricted (binary) terms, being an un-masculine man can lead to discrimination and violence. In countries such as Australia, homosexual men suffered a long history of such discrimination. Openly gay, effeminate men were easy targets throughout the 20th century, although the rise of glam-rock in the 1970s and the metrosexual man in the 1980s broadened the spectrum of masculinity in the Western world. For some, this was the beginning of the masculinity crisis in the West. Joe Jackson's 1989 hit song, 'Real Men' exemplified the crisis, mourning as it did the simpler days when girls wore pink and boys wore blue.

The existence of this crisis has been questioned, however. Beynon (2002) posited that much of the crisis talk was fabricated by journalists and that men were not preoccupied with how to be a man. Many authors, he claimed, have pointed to the continuance of power in the hands of the heterosexual man. Even so, Plain (2000, cited by Beynon, 2002) argued that the masculinity crisis could be observed in the disappearance of the hard-boiled detective from books and Hollywood screens, just as Barry (2017) complained about the sensitive, new-age James Bond portrayals.

Masculinity in Western countries such as Britain, the United States and Australia are, therefore, shaped to some degree by conflicting portrayals in popular culture. There are more ways to be a man and, therefore, to be masculine, than the traditional movie stereotype apparently allowed for (Vanderbeck, 2005). Yet masculinity is still discussed within the framework of hegemony, with variances around the accepted norm seen as somehow deviant or deficient (Barry, 2017). Such masculinities are rather more amorphous than the term 'hegemony' would imply. The pop-culture hegemonic portrayals, too, are prone to evolution (Atkinson, 2012). Yet there is evidence that while expressed attitudes can change rapidly, behaviourally expressed masculinity tends to be more inert. Lass (2018) suggests that the resistance to changing behaviour is economically based, since men generally enjoy an earnings advantage.

Biology appears to have played a vital role in determining the division of labour in traditional societies, with men usually taking on the roles of hunter and warrior and women those of child bearer and primary carer. Darwin's sexual selection theory further posited that reproductive preferences might contribute to some of the physical attributes that dictated this division of labour (Jones & Ratterman, 2009). More recently, researchers have expanded the concept of sexual selection to include non-physical traits such as behaviours and attitudes, preferences for which might be determined individually or collectively and socially, rather than genetically (Janicke & Morrow, 2019; Nakahashi, 2017). Hence, it is probable that masculinity (how one displays one's maleness) is influenced in many societies by the preferences of women. This is evidenced, for example, by the stick fighting rituals common to many sub-Saharan African tribes. The much publicised *sagine* of the Surma tribe of Ethiopia is in no small part aimed at demonstrating courage and strength to prospective wives. Marriageable girls observe the fights, choosing to offer themselves to fighters who display the desired attributes (Aniago, 2019). Similarly, a Samburu warrior who kills a lion during the initiation hunt is likely to impress prospective wives with his masculine qualities. According to one Turkana chief, 'It is the young women, the girls of our tribe, who incite the young men to deeds of bravery' (Lytton, 1957, quoted in Huxley, 1997: 325).

It is, of course, notable that sexual selection/preferencing is not the exclusive domain of females. Both genders might impose such selection (Jones & Ratterman, 2009), raising the probability that femininity is also influenced strongly by male selection. In Kenya, changing femininity has been observed in some Maasai communities, where warriors have refused to marry circumcised girls. Where once a girl was not considered a marriageable woman until she was circumcised, now women are choosing not to circumcise their daughters in order to improve their marriage prospects. While men have not been the sole actors in the anti-circumcision movement, they have been a vital component in the shifting femininity (Onyulo, 2016).

As important as sexual preference might be in defining masculinity/femininity, other factors are also involved. Silberschmidt (2004) pointed to the role of colonisation in creating a crisis of masculinity among the

Gusii people of Western Kenya. The British brought industrialisation to Kenya, a socioeconomic system for which the Gusii were ill-equipped. As warriors and farmers, the Gusii did not possess business skills or education that would enable them to participate profitably in the new economic world. They were relegated to the life of itinerant workers and the marital infidelity with which it is associated. The Gusii women meanwhile took over the family farms, freeing them from the traditional sexual-economic cycle of dependency (Gentry, 2007). Faced with this changing femininity, and women who regarded them as useless, Gusii men fell into a crisis of masculinity to which they responded with increased incidence of domestic violence and infidelity (Silberschmidt, 2004). Gusii masculinity was redefined by the changing economic system imposed by colonisation and changing femininity as well as expectations from their women that they felt incapable of matching. Hence, multiple factors, including women's expectations, interacted to form the new masculinity.

Although Silberschmidt (2005) contends that male disempowerment has shaped the emergent masculinity more broadly throughout East Africa, significant cultural differences result in varied masculinities between communities. The country's 44 official tribes (Roychowdhurry, 2017) are classified according to ethnolinguistic origins, with three main indigenous groups of Nilots, Bantus and Cushites. Significant cultural variations, including expressions of masculinity and femininity, can be observed, not only between ethnolinguistic groups but also within them. For example, some Nilots, such as the Maasai, traditionally practise circumcision for both men and women as a central component of the coming-of-age initiation. Others, such as the Luo, do not have a circumcision tradition, preferring to remove two front teeth, although with the prevalence of AIDS, Luo men are now choosing to be circumcised as a preventative measure, demonstrating a medical influence on their masculinity. Another Nilotic tribe, the Turkana, do not initiate men or women into adulthood with any type of ceremony or physical mutilation. The Turkana appear to have experienced no masculinity crisis, perhaps because they occupy arid lands that the British regarded as valueless. Consequently, colonisation had little impact on them (Makoloo, 2005).

In addition to the three indigenous ethnolinguistic groups, three 'foreign' tribes are also recognised. These are Europeans (*Wazungu*), Indians (*Wahindi*) and Arabs (*Waarabu*). The masculinities vary noticeably between these three tribes and even within them. The *Wahindi*, for example, are not a homogeneous group and some Indian families are matriarchal. Others have patriarchal traditions, but it has been observed that diaspora Indian communities are becoming more gender equal, particularly where the wife is involved in family businesses or otherwise employed (Lobo, 2018; Sonawat, 2001).

Even so, there are some common attributes of men and women throughout Kenya. The act of marriage and reproduction is a key feature of adulthood in Kenya. Hence it defines both masculinity and femininity. One is not truly accepted as an adult man or woman until one has children. Being childless, gays and lesbians are considered to be at odds with society (Patron, 1995) and there is a high level of homophobia in the country. The production of children also dictates gender roles, as already discussed. The men are expected to engage in dangerous physical activities while the women adopt more passive, nurturing roles. Men are also regarded, in most communities, as the heads of the household, and are recipients of inheritance and land ownership benefits as a result. In some communities, such as the Nandi and the Gusii, older women who have never married and who have no children are able to marry one or more female wives who bear children on their behalf, guaranteeing continuance of the family line. Such women-marriage arrangements are plutonic, but they confer the rights, status and responsibilities of manhood upon the female *husband* (Nyanungo, 2014).

Masculinity is broadly defined by the role of provider, protector and dominant partner in both marriage and society (see Silberschmidt, 2004). Women-marriages notwithstanding, physicality is important too. The true man should look like a man. This aspect of Kenyan and, more broadly, sub-Saharan African masculinity has been illustrated through several personal encounters. One Kikuyu member of the Kenyan diaspora in Melbourne was referred to by a Congolese friend of mine as 'that man who looks like a woman'. The Kikuyu man's soft features were not well regarded and despite his being a father to three boys, he was not considered to be entirely a man. Similarly, a young Kikuyu man that I met through research and maintained a friendship with confided that he was bullied at school, and even in adult life, because people said that he looked like a woman. Again, his soft facial features were not sufficiently masculine, according to his fellow Africans.

A certain level of bravado is arguably necessary for one to be truly regarded as a man in Kenya. I was once derided by a Kenyan man because I walked 'too softly'. I looked weak and insignificant, he said. He then pointed to a large white woman who was strutting around, looking extremely confident. 'She is manlier than you', he told me. The notion that one's *presence* is a marker of masculinity was further illustrated by a Gusii friend's regret at having lost weight through illness. He was once a fat man and he expressed a wish that he could once more grow fat because when a fat man enters a room, everybody takes notice. I have also been greeted by Gusii women with arguably the greatest compliment affordable to a man, saying that I 'look strong'. True masculinity involves having a physical presence, which is at least partly created by size, and by being noticed. In other words, importance is a marker of masculinity.

While varied to some extent, a masculine hegemony is apparent in Australia and other Western countries. The picture, however, is more complex in Kenya. Despite some generally accepted masculine traits, linked in no small part to the reproductive function of an adult, tribal diversity has resulted in significant variations in how masculinity is displayed. Differing circumstances following colonisation have exacerbated inter-tribal variations in masculinity and femininity. It is, therefore, difficult to claim that a masculine hegemony exists in Kenya.

Masculinity in the Field, Kenya

Despite having confronted issues of masculinity, in both my personal encounters and my research in African contexts, this is not something that has invaded my consciousness as a researcher to any great extent. Nevertheless, it is apparent that there are three elements upon which perceptions of masculinity can act in field research. Dissonant masculinity can have an impact on the research itself, potentially altering the trust relationship between researcher and participant. It can also have an impact on the community being researched, causing discomfort or concern. Finally, dissonant masculinity can affect the researcher himself, possibly causing distress and feelings of inadequacy or disconnection (Vanderbeck, 2005). Conversely, harmonious masculinities might be regarded as having an opposite effect.

Dissonant masculinity and trust (impacts on the research)

It is easy to imagine that presenting a non-hegemonic masculinity to a research participant or community might result in a failure of trust. After all, manliness is important in Kenyan and other African societies. Although my support for gay rights has led to some Kenyans labelling me as a homosexual, such discussions have only occurred in social settings unrelated to research. The logic of such name calling has been easily dealt with by pointing out that I am not a woman but I support women's rights, and I am not black but I support black rights. However, I have not tested the effectiveness of such a response in the research field because the topic of gay rights has never arisen. There is good reason to believe that creating an impression of being gay would affect trust with most Kenyan and sub-Saharan people, but this is something that I, as a heterosexual man, can only speculate about. It is my philosophy that the researcher is not in the field to change opinions; rather he is there to document and analyse them. Consequently, my thoughts on homosexuality and gay rights have never been aired during research.

While a masculine presentation might be important when conducting research in Kenya, some less masculine physical traits cannot be altered. A Luo woman once told me that my hands were those of a woman, not a man. However, this was in a social context and I had known her for many years. She only noticed the femininity of my hands because I had drawn a logo for which I used my own hands as models. She asked why I had drawn a woman's hands and then determined that my hands were not masculine. If not for the artwork, she would not have paid any attention to my slender fingers. Similarly, no research participant has ever commented on my hands. Facial features are more apparent, however. As a bearded man, my facial masculinity has not been questioned. As mentioned earlier, my lack of presence (i.e. physical size) has been commented on, but never in the context of research. It is conceivable, nonetheless, that it has been noted by my participants. Whether or not it has been, I have detected no effect on my research. I have not been treated with suspicion and I have collected thick data in every study.

In addition, I am a softly spoken man, and this has been commented on. Africans tend to be quite loud, especially in social gatherings. But this is not a strictly masculine trait. Women also tend to speak at volume in social settings. These characteristics present a somewhat mixed masculinity to Kenyan people. I am not effeminate, but I do not display the bravado and attention-seeking attributes that arguably typify the African man. In addition, until recently, I did not have a family and was therefore regarded as not really a man. This was exacerbated during my PhD research because, as an elder, it was considered strange that I should still be in school. One Maasai elder with whom I had a long discussion was particularly adamant that I was too old to be at school. He also recommended that I marry a local woman and move with her to Mombasa, since that was my birthplace.

Yet these dissident presentations of masculinity did not seem adversely to affect the trust relationship between my participants and myself. While people such as the Maasai elder were prepared to give me advice on how to be a true man, such advice was given gently and did not create awkwardness or bring about an end to the conversation. Masculinity does not operate in isolation from other social aspects. Having been born in Kenya, I was immediately accepted as a fellow Kenyan. My understanding of the cultural landscape also ingratiated me to most people with whom I interacted. My willingness to eat the same food that Africans eat was another trust builder, in both Kenya and Botswana. Similarly, my love of African music and desire to hear local variations conveyed an alignment with African culture. The recognition that I was of a different tribe (a *Mzungu*) also allowed me to present with some degree of *reduced* masculinity. Nonetheless, I still bemused many people because my definition of being a man was not entirely in sync with theirs.

Among the Gusii in Western Kenya, for example, despite identifying as married and having children during my most recent research, my masculinity was still questioned on occasion. One huge man, a local politician, introduced me to a young woman, suggesting that I have sex with her. The Gusii, in general, measure their masculinity in no small part by the number of women with whom they engage sexually (see Silberschmidt, 2004). My response, that my wife would not approve, was met with the explanation that Gusii men prefer 'young girls like this one'. The disparate definition of what it means to be a man was quickly resolved by stating the obvious, 'but I am not a Gusii'. This was met with laughter and acceptance of my *Mzungu* ways. Perhaps, strangely, the identifiable differences did not hamper my immersion into the community. Kenya's ethnic diversity facilitates acceptance of different ways, especially when the one displaying the difference is from a different tribe (see Lacey, 2017).

While there is potential for eccentric masculinity to create issues, depending on the level of the disparity and other interacting social forces, it can also create opportunities. My PhD research investigated the phenomenon of philanthropic tourism to a Kenyan charitable field project in a district dominated by Kikuyu people (see Figure 10.3). Unexpectedly, it uncovered instances of sexual abuse of children, and violence perpetrated by management towards staff and local people. Such activities were conducted only by male managers and staff, and arguably represent quite common expressions of Kenyan, and more generally African, masculinity (Adamson, 2017). Sexual abuse of school children by teachers or other school staff is common throughout the country, despite being illegal (Lacey, 2017). While there are many honourable, law-abiding teachers, and men more generally, Kenyan masculine displays are frequently tied to sexuality, which is often exploitative or abusive. Women are commonly reluctant to trust men with reports of such crimes or even to confide in them for a cathartic release. Several women, however, opened up to me, with one actively seeking me out so she could tell me her story. Our interview was lengthy and deeply emotional. The woman cried through much of it and ended by thanking me for enabling her to get the things she had witnessed off her chest. She had sought me out not because I was a *Mzungu* who

Figure 10.3 The PhD study site, a school and orphanage in the Gatarakwa sub-district, Nyeri, Central Kenya

Source: Author.

might be able to do something about the issues but, she said, because she could not speak to anyone else. My *foreign* identity combined with my eccentric masculinity created trust and presented me with data that I might otherwise not have uncovered. In addition, it provided the woman with an emotional release. Had I presented similarly to a typical Kikuyu man, it is doubtful that such a level of trust would have been established. The outsider-friend (one who is not born to a community but who has the community's interests at heart, who is regarded by the community as a friend and whose culture is at variance, to some degree) can often achieve what someone who displays complete cultural similarity cannot. This is the case with non-hegemonic masculinity and femininity alike (see Lacey, 2017).

Impacts of dissonant masculinity on the community

While there is undoubtedly potential for non-hegemonic masculinity to cause distress to the host community, I have not been aware of any significant issues in my own field research or even social interactions in Africa. This is perhaps because my masculinity displays are eccentric, but not *deviant*. That is to say, I do not display identical masculinity to the African communities in which I have been immersed, but my variation on masculinity is not so offensive as to hinder being trusted or welcomed into the community. I was identified as unmarried for much of my research and I was also identified as still being at school despite being in my 50s at the time of my PhD study. I am not promiscuous, and I relate to women in a different manner from the average African man. I cook and I do my own washing and cleaning. While single men often undertake their own domestic chores, one with sufficient wealth is expected to outsource the laundry. More unusually, I continue to cook and to do my own washing as a married man. Washing clothes is seen as women's work and my laundry duties are conducted even over the objections of my wife and daughter who absolutely forbid me to do their laundry, even when they are unwell. Hence my display of masculinity has been modified a little by the preferences of the women in my life. It is notable, too, that I am not permitted, under any circumstances, to launder the clothes of a younger, unmarried man who lives at our house. As a Gusii man, he tells me this is absolutely against his culture. Masculinity is linked to age, wealth and marital status and differing behaviours are sometimes expected according to these factors. Furthermore, the taboo on my washing my wife's and daughter's clothes is linked to their femininity. They fear being judged, by other women, as not looking after their husband/father properly.

The above-mentioned characteristics are at odds with the masculine and feminine norms of the specific communities with which I mostly interact. On the other hand, my masculinity still conforms to some key normative rules. I am heterosexual, and I engage in traditionally masculine pursuits such as martial arts, welding and physical labour. Moreover, I am comfortable around cattle (see Figure 10.2), having lived on a smallholding with livestock for part of my childhood. Being bald (many Kenyan men shave their heads) and bearded (most of the time), my facial appearance is sufficiently manly.

In a multicultural country such as Kenya, in which masculinity is displayed with a degree of variation between ethnic groups, my eccentric masculinity is accepted as that of a *Mzungu*, or at least as one who is not of the same tribal identity as my research participants. Notably, I have been able to immerse myself in Kenyan communities such as the Kikuyu of Nyeri district and the Gusii from Kisii, within a week or two of arrival. During my PhD research in Gatarakwa subdistrict, Nyeri, I was so accepted as a member of the community that even a year after returning to Australia, I received a message requesting that I appoint a delegate to represent me at a community meeting. Any variances in my masculinity, such as being unmarried and childless, were apparently outweighed by my identity as a Kenyan and the recognition that, as a *Mzungu*, I did not have to conform perfectly to Kikuyu norms.

The variations in masculinity or femininity that are allowable are often greater for members of other communities. This was highlighted by the subject of my research into HIV education, who was able to talk about sex and sexual health to her own Gusii community only by pretending to be a Kikuyu woman, and therefore not being bound by the norms applied to Gusii women (see Lacey, 2017). Mayaka (2015) has highlighted the importance of the identity as an outsider/insider friend in gaining trust in four different communities in Kenya, demonstrating the willingness of Kenyans to accept different cultural attributes from those of another tribe or ethnicity. Since these concessions are made because of differing tribal identities, my dissonant masculinity would likely be less tolerable if I were of the same ethnicity as those to whom it is displayed.

It is possible, too, that more irritation might be shown were I not seen as accepting of other cultural practices and beliefs. My interest in hearing and voicing community concerns, in sharing meals in people's homes and in engaging as a member of the community undoubtedly enhanced my acceptance in the community. Being a Kenyan but of a different tribal identity also placed me in a unique position of trust, since I was not seen to have an agenda in the community and was not regarded as a foreigner seeking to take advantage of the community (see Lacey, 2017; Mayaka, 2015; Mayaka *et al.*, 2019). Hence, my masculinity, while at odds with normative values, was accepted with little more than bemusement. It was shrugged off as that of a *Mzungu*. Although the notion appears to have received little direct attention in the literature, it seems apparent that masculinity can be regarded as relative, and as interactive with other social markers (see Miller, 2016). As long as it does not stray too far from the norms and is accompanied by other culturally acceptable attributes, unconventional masculinity does not appear to cause any great offence to Kenyans.

Impacts of dissonant masculinity on the researcher

As is the case with many cultural practices and displays, dissonant masculinity has the potential to create discomfort and even fear in the researcher. Dissonant cultures commonly produce symptoms of culture shock in travellers, a condition Mills (2007: 80) describes as resulting from 'an excess of novelty' (see Figure 10.4). While culture shock is usually short lived, Mills states that it can become severe enough to cause stress-related illnesses and create distress, crying and premature cessation of the journey. In the case of culture shock resulting from rejection of one's masculinity, there could be accompanying fear of attack. This is likely to be especially true in Kenya if the researcher is perceived as gay. According to Mills (2007), culture shock can also result in feelings of loneliness and isolation. Hence, any cultural differences that limit acceptance of the researcher into the host community are likely to increase the likelihood of culture shock. Fortunately, despite differing masculinities and other cultural attributes, including religion and even the way people drive, my Kenyan birth and close association with Africa (through the African diaspora of Melbourne and family connections in Kenya) have shielded me from especially troublesome culture shock. My understanding of African culture has allowed me to slip into African life in both Kenya and Botswana with little trouble. Being aware of, and therefore prepared for, African masculinity has equipped me for comments and attitudes that might otherwise have caused distress. While I do not make overt efforts to display a more African masculinity, I am aware of subtle changes in the way I walk and interact when in Africa. If for no other reason than safety, I have found it necessary to adopt a more confident posture and general body language when in Kenya. It is also necessary for me to be more

Figure 10.4 The outskirts of Kisii Town in Western Kenya; it is in such areas and rural settings that poverty and culture shock are most likely to be encountered
Source: Author.

assertive in group conversations. Having a presence is important in Kenya. Without some assertiveness a person goes unrespected, and this seems to be especially true for a man. However, I have not found such assertiveness to be particularly important in research contexts. A degree of confidence is required, particularly when conducting research in potentially dangerous settings such as side streets in towns, but I have not found it necessary to be assertive or to talk over people in the course of conducting interviews, for example, since my function as a researcher is to listen.

Culture shock is likely to be far more severe, and potentially longer lasting, for a researcher who has less connection with Kenya or Africa more broadly. Adverse community responses to a researcher's masculinity will likely be exacerbated by additional cultural variations. Lack of religion, unfamiliarity with the national or the local language, misunderstandings arising from the Kenyan dialect of English, lack of willingness to eat local food, etc., all interact with one another and lead to a greater sense of being an outsider. In my case, the main differences are to be found in masculinity and religion. In other cultural areas, the issues are less pronounced (although I readily concede that improvement in my Kiswahili language skills would be helpful). I am comfortable with eating Kenyan food and often cook it at home in Australia. I frequently carry a stick that resembles a Maasai *rungu* (see Figure 10.5) which immediately endears me to Maasai people, and

Figure 10.5 The author with a hiking stick (behind), gifted by his father, and a Maasai *rungu* that it resembles; for the Maasai, the *rungu* is a key cultural artefact and symbol of masculinity which is carried at all times
Source: Author.

I generally follow Kenyan social protocols, including the manner of greeting people. These, combined with my *Mzungu* appearance, my having been born in Kenya and owning a house in the country, allow some leeway in both masculinity and religion, although it is occasionally necessary for me to remind those with whom I engage that I am a *Mzungu*, not an African. Immersion in a community does not necessarily imply trying to be accepted as homogeneous with that community, as demonstrated by the colonial memoires of Mirella Ricciardi (Huxley, 1997).

Conclusion

In Kenya, and most probably throughout sub-Saharan Africa, masculinity is a dependent variable that is judged in combination with other social or cultural indicators. An eccentric masculinity displayed by a *Mzungu* is likely to be less disruptive if accompanied by other signals of alignment to Kenyan culture or identity. Understanding social protocols, a willingness to eat Kenyan food, having respect for religious beliefs (even if being at odds with them), listening to Kenyan or other African music, especially that which is specific to the region in which one is conducting research, and speaking the national or local language all figure into the equation of acceptance. Being of mixed race and having been exposed to cross-cultural elements in my childhood have facilitated my acceptance of and empathy for differing cultures (see Fopp, 2008). Being identified as a fellow Kenyan has also helped me in my research. Acceptance as a friend allows for significant, but not unlimited, cultural variance, including the definition and display of masculinity.

While it might be assumed that alignment of masculinity between researcher and study community would afford greater acceptance and trust, my changing status from single and childless to husband and father, and therefore as closer to the Kenyan definition of a man, has had no noticeable impact on my research or my social interactions. This underlines the degree to which my dissonant masculinity has always been tolerated by my Kenyan interlocutors. That tolerance is undoubtedly born of the cultural diversity of the country as well as my own cultural breadth.

References

Adamson, A. (2017) Divergent masculinity discourses among Stellenbosch student males: Traditional masculinities and the progressive male/new man discourse. MA thesis, University of Stollenbosch. See https://scholar.sun.ac.za.

Aniago, E. (2019) Thick description of social functions of selected African flogging-bouts as theatrical entertainment and self-defence martial arts. *Ido Movement for Culture. Journal of Martial Arts Anthropology* 19 (1), 9–19.

Atkinson, N. (2012) James Bonds I used to know. Every generation gets the James Bond it deserves. *National Post*, 7 November. See http://www.nationalpost.com/entertainment/nathalie-atkinson-james-bonds-i-used-to-know/ (accessed 10 August 2018).

Barry, J. (2017) Why Ryan Gosling should never, *ever* be James Bond. *Flare*, 2 March. See https://www.flare.com/celebrity/ryan-gosling-james-bond/ (accessed 10 August 2018).

Beynon, J. (2002) *Masculinities and Culture*. Buckingham and Philadelphia, PA: Open University Press.

Connell, R.W. (2005) Globalization, imperialism, and masculinities. In M.S. Kimmel, J. Hearn and R.W. Connell (eds) *Handbook of Studies on Men and Masculinities* (pp. 71–89). Thousand Oaks, CA: Sage.

Couric, K. (2017) *Gender Revolution*. National Geographic DVD.

Fopp, R. (2008) *Enhancing Understanding, Advancing Dialogue: Approaching Cross-Cultural Communication*. Hindmarsh: ATF Press.

Forth, C.E. (2008) *Masculinity in the Modern World*. Houndmills: Palgrave MacMillan.

Gentry, K.M. (2007) Belizean women and tourism work: Opportunity or impediment? *Annals of Tourism Research* 34 (2), 477–496.

Huxley, E. (1997) *Nine Faces of Kenya*. London: Harville Press.

Janicke, T. and Morrow, E.H. (2019) Sexual selection. *Evolution, Medicine and Public Health* 2019 (1), 36.

Jones, A.G. and Ratterman, N.L. (2009) Mate choice and sexual selection: What have we learned since Darwin? *PNAS* 106 (1), 10001–10008.

Lacey, G. (2017) Delivering culturally sensitive sexual health education in western Kenya: A phenomenological case study. *African Journal of AIDS Research* 16 (3), 193–202.

Lass, I. (2018) The division of paid and unpaid work among couples. In R. Wilkins and I. Lass (eds) *The Household and Labour Dynamics in Australia Survey: Selected Findings from Waves 1–16. 13th Annual Report of the HILDA Survey*. Melbourne: Melbourne Institute for Applied Economic and Social Research.

Lobo, A. (2018) Anglo-Indian Women: A narrative of matriarchy in a global diaspora. In A. Pande (ed.) *Women in the Indian Diaspora: Historical Narratives and Contemporary Challenges:* Singapore: Springer Nature.

Makoloo, M.O. (2005) *Kenya: Minorities, Indigenous Peoples and Ethnic Diversity*. London: Minority Rights Group International and Cemiride.

Mayaka, M.A. (2015) The role of entrepreneurship in community-based tourism. Doctoral dissertation, Monash University.

Mayaka, M., Lacey, G. and Rogerson, C.M. (2019) Empowerment through friendship: A process view of community-based tourism. Presentation to the Proceeds of Travel and Tourism Research Association Advancing Tourism Research Globally Conference, Melbourne.

Miller, D. (2016) Intersectionality: How gender intersects with other social identities to shape bias. *The Conversation*, 4 February. See https://theconversation.com/intersectionality-how-gender-intersects-with-other-social-identifiers-to-shape-bias-53724 (accessed 10 September 2017).

Mills, D. (2007) *Travelling Well* (14th edn). Brisbane: Self-published.

Nakahashi, W. (2017) Cultural sexual selection in monogamous human populations. *Royal Society Open Science*, 21 June. See https://royalsocietypublishing.org/doi/full/10.1098/rsos.160946.

Nyanungo, H. (2014) *Female Husbands without Male Wives: Women, Culture and Marriage in Africa*. Johannesburg: Open Society Initiative for Southern Africa (OSISA). See http://www.osisa.org/buwa/regional/female-husbands-without-male-wives-women-culture-and-marriage-africa (accessed 12 August 2018).

Onyulo, T. (2016) Alternative to genital mutilation emerges for Kenya's Maasai girls. *Newsweek*, 26 March.

Patron, E.J. (1995) Heart of lavender: In search of gay Africa. *Harvard Gay and Lesbian Review*, Fall. See https://sourcebooks.fordham.edu/pwh/patron-africhomo.asp (accessed 11 August 2018).

Roychowdhurry, A. (2017) Indians become the 44th tribe of Kenya; but what does that mean? *The Indian Express*, 25 July. See https://indianexpress.com/article/research/indians-become-the-44th-tribe-of-kenya-but-what-does-that-mean-4766526/ (accessed 18 August 2018).

Silberschmidt, M. (2004) Men, male sexuality and HIV/AIDS: Reflections from studies in rural and urban East Africa. *Transformation: Critical Perspectives of Southern Africa* 54, 42–58. doi:10.1353/trn.2004.0026

Silberschmidt, M. (2005) Poverty, male disempowerment, and male sexuality: Rethinking men and masculinities in rural and urban East Africa. In L. Ouzgae and R. Morrel (eds) *African Masculinities* (pp. 189–204). New York: Palgrave MacMillan.

Sonawat, R. (2001) Understanding families in India: A reflection of societal changes. *Psicologia: Teoria e Pesqisa* 17 (2), 177–186.

Tedlock, B. (1991) From participant observation to the observation of the participation: The emergence of narrative ethnography. *Journal of Anthropological Research* 14 (1), 69–94.

Vanderbeck, R.M. (2005) Masculinities and fieldwork: Widening the discussion. *Gender, Place & Culture* 12 (4), 387–402.

11 Doing Fieldwork in Palestine: Checkpoints, Access Restrictions, Security and Well-being

Rami K. Isaac

Compared to past times, many people now have the means and opportunity to travel easily and freely. However, freedom of travel and the right to travel are far from being globally recognised as human rights. For some travellers, freedom of movement is determined by their race, nationality and, particularly, their gender. Drawing on a personal account of undertaking fieldwork in the sociopolitical context of Palestine, I use this chapter to explore my positionality as a male researcher, and how this influences my personal security, freedoms of access and movement restrictions, as well as my well-being.

Introduction

Critical tourism scholars have expressed concern about the systemic structures (Cole & Eriksson, 2010; Higgins-Desbiolles, 2006) and inequitable practices (Hall, 2010; Hall & Brown, 2010) that violate individual freedom of movement and the right to travel. In spite of the United Nations Declaration of Human Rights, individual freedom of movement is far from being universally accessible and remains a highly ambiguous and politically charged matter (Bianchi & Stephenson, 2013). In the name of safety and security, governmental changes have been made both within and across borders internationally to intensify surveillance and gather data for policing and intelligence (Bianchi & Stephenson, 2013; Lyon, 2003). The aim of this chapter is to explore my fieldwork experiences vis-à-vis my positionality as a male researcher, and how this influences my personal security, access to different regions in the West Bank and study participants, freedom of movement and well-being.

The chapter will start with a discussion of politics and emotions, and then move on to a reflection on the debates over geographies of fear, gender, well-being and place. The chapter introduces Palestine and outlines three important impediments in Palestine, namely the security and well-being of Palestinian researchers, the im/mobilities of Palestinian male researchers and the Checkpoint 300 Gilo terminal in Bethlehem. To the best of the author's knowledge, no study has yet been published with a focus on the experience of a male Palestinian researcher in Palestine. The intention of this chapter is to help close this gap in the literature, and by doing so to increase understanding of the complex politics of masculine emotion and fear caused by movement restrictions and access restrictions.

Politics and Emotion

Drawing on the work of Connell (2005), who has theorised various relations of power between masculinities and feminities, an emphasis on politics, particularly in the case of Palestine/Israel, is important for research on emotional geographies for a number of reasons (Meth, 2009). Constructions of 'excessive emotionality' (fear, irrationality) are linked with ideas of powerlessness (Meth & Bondi, 1999: 75) and have, in Western contexts, historically informed what it means to be feminine (Sharp, 2009). Furthermore, authors Sharp (2009) and Bondi (2005) have argued that there is a potential danger in the ways in which feminist geographers have approached questions of emotion. They claim that individualised accounts of emotion may neglect a broader 'collective politics' and a more critical questioning of subjective accounts (Bondi, 2005; Sharp, 2009). As Meth (2009: 854) noted, 'I would encourage an extension of this call for politics which incorporates marginalised men, but one that operates with sensitivity to the political implications of men's claim over gendered power relations'.

Several levels of intersection between emotion and politics have been identified within a political science framework (Meth, 2009):

- emotions which are 'part of the human conditions' (basic emotion);
- emotions occurring at the 'epochal level' (such as Apartheid) and which characterise an age in history;
- emotions which are particular to the experiences of a social group and are experienced as 'abiding affects';
- emotions as strategically organised responses to a particular political situation; and finally
- emotions as fleeting reactions to everyday political practices (see Clarke et al., 2006: 11–12), such as the case of Palestinian citizens' emotional reactions to the Israeli occupation.

Emotions matter, yet they have been noticeably absent from previous tourism research. Jamal and Hollinshead (2001: 67) argued that '[t]he omission of studies and narratives which locate … "emotion" in tourism, whether that of tourist or the host, is a problem which has been noted and addressed by very few scholars'. As Buda (2012: 52) states, 'their call for more recognition of emotion in tourism studies a decade ago seems to have been a cry that remains unheard'. Different accounts of shame and pride (Johnston, 2005, 2007; Tucker, 2009; Waitt *et al.*, 2007) and fear (Mura, 2010) in tourism have recently been published. Nevertheless, debates on the place of emotions in tourism research remain largely lacking. Questionably, the topic has been marginalised through the gender politics of research wherein the tourism academy is conditioned to principles of distance, objectivity and rationality (McIntosh, 2010). The following framework is a starting point from within which the intersection of political practice and men's emotions can be explored.

Geographies of Fear, Gender and Place

Meth and Bondi (1999: 77) describe the ways in which men 'take for granted their capacity to move through a variety of urban spaces'. Nevertheless, recognition of men's experiences of fear is now evident in work that argues that geographies of fear overlap with difference in terms of gender, race, age and sexuality (see Brownlow, 2005; Day, 2006; Pain, 2001, 2003). Certain men are very vulnerable to violence. Whereas some men openly express fear, this is often circumstanced and designed by context and possible cultural factors, as well as a strategic understanding of the risks of appearing vulnerable (Brownlow, 2005: 589). Brownlow (2005: 590) states that the 'suppression of fear does not translate into the termination of fear'. This recent work demonstrates the intersection of place, gender and marginalisation and brings attention to the differentiated, fearful emotions of men as well as the politics of fear (Shirlow & Pain, 2003). Interpretations of men's fear can also refer to fear of humiliation (as in the case of Palestine under occupation) and social stigmatisation.

Geographical understandings of men's emotions in non-Western contexts are particularly limited (Meth, 2009). Therefore, in spite of increasing research on emotional geographies (see Davidson *et al.*, 2005; Sharp, 2009), there is relatively little research that uses a gendered lens to analyse the wider emotional features experienced by marginalised men. This chapter uses Palestine as a case study, drawing on my personal accounts and field experiences. Meth (2009: 855) points out the absence of research into emotions such as a 'feeling of inferiority, anger and uselessness'. There exists, in addition, little research that examines the politics of emotion in relation to how particular power practices are experienced by men, as in the case of Palestine, which is effectively controlled by an external power – Israel – with consequent restrictions on access to and mobility within Palestine.

Reflexivity in Tourism

Reflexivity is the acknowledgement of the agency of researchers, those being researched, scholars and others in producing knowledge (Tribe &

Liburd, 2016). Reflexivity is increasingly seen as being crucial for researchers across the social science disciplines. The need to account for the position of the researcher leads to an increased focus on the unique worldview with which the researcher interprets what they encounter during the research process. By incorporating and interrogating their own positions in both their research and writing, social scientists are able to create more rounded empirical accounts of the social world. Reflexivity thus enables the researcher to acknowledge their own situated role and to weave their voice through the research process. This chapter is written from an outsider/insider's standpoint: that of a Palestinian tourism researcher conducting fieldwork in Palestine. Along with Crang (2003: 496), I am weary of 'work that divides positionality formulaically into being insiders (good but impossible) and outsiders (bad but inevitable)'. As I am familiar with the Palestinian region, I attempted to 'write myself into' the research process, because one avenue encouraging reflexivity is the 'writing-IN of qualitative research ... by the use of personal pronouns (for example, I, we, our)' (Mansvelt & Berg, 2005: 257). My own positionality in the field as Christian Orthodox, Palestinian and a young man somewhat strengthens the powerful researcher–researched dichotomy in the research process (Mansvelt & Berg, 2005).

Palestine

Palestine is a unique, faith-associated tourist destination; its long history, religious significance and natural beauty make it an amazing place to visit. Palestine's significance derives partly from the fact that it is home to the three monotheistic and Abrahamic religions of Judaism, Christianity and Islam. Every year Palestine attracts many pilgrims who visit the holy places. Nevertheless, since the beginning of the 20th century, Palestine has seen complicated changes in its political and social circumstances (Isaac et al., 2016). One of these changes was the creation of the State of Israel in 1948. Subsequently, in 1967, Israel occupied the Palestinian West Bank, including East Jerusalem and the Gaza Strip.

In spite of the signing of the Oslo Accords between Israel and the Palestinian Liberation Organisation (PLO) in the 1990s, and the acceptance of the establishment of the Palestinian Authority to administer some parts of the Palestinian territories, which are merely major urban centres in Palestine, many such areas remain under Israeli occupation and control. Israel controls all access to Palestinian cities and towns, most Palestinian water resources and all movement of people and goods from, to and within Palestine (Isaac et al., 2016). One of the most important geographic implications of the Oslo Accords is the designation of Palestinian land into Areas A, B and C, representing the extents of Israeli and Palestinian jurisdiction and the policy of closures (Tawil-Souri, 2011). Many scholars (see Harker, 2009; Isaac, 2013; Weizman, 2007) report that numerous

Palestinians living in the West Bank and the Gaza Strip face considerable obstacles to their mobility when they attempt to travel within Palestine. The lack of control over their borders, their vulnerability to regular incursion and subsequent physical damage to tourism infrastructure, the lack of freedom of movement for Palestinians and tourists, the regular closures

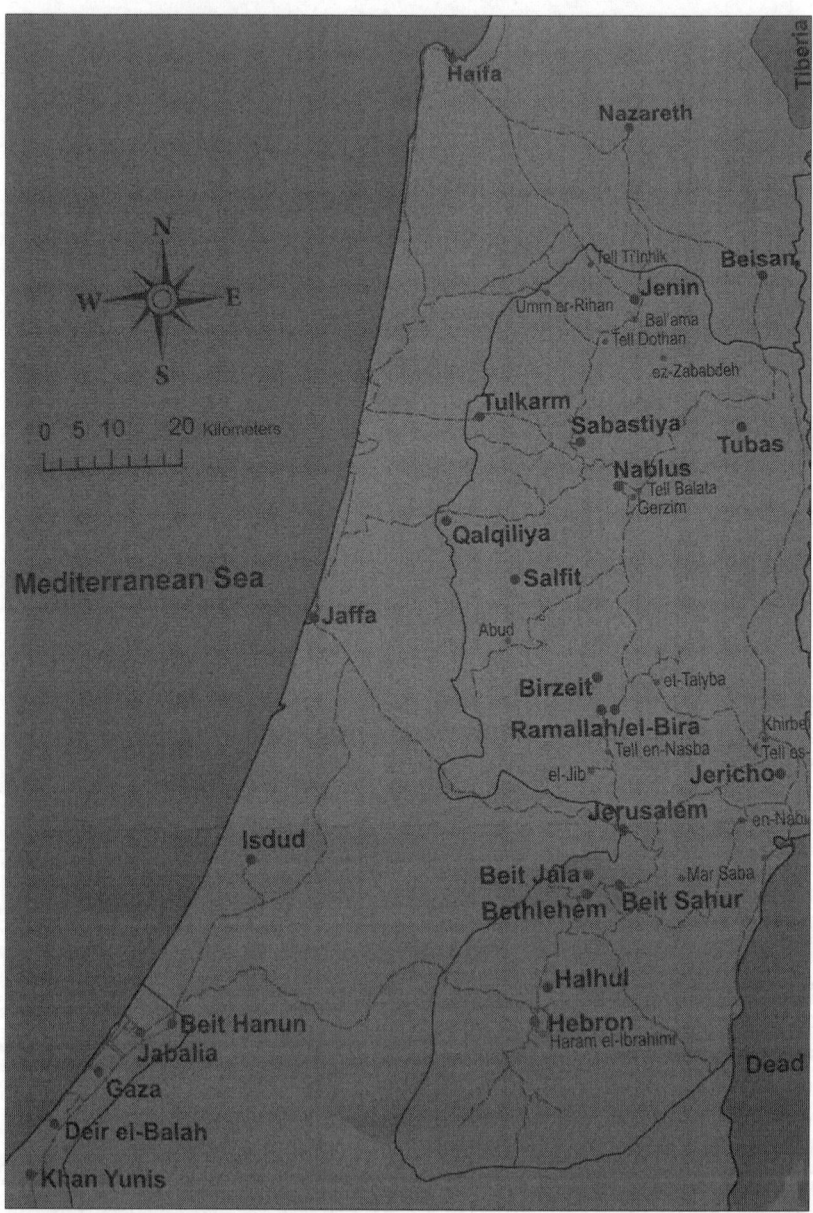

Figure 11.1 Palestine

of Palestinian areas, and the Segregation Wall (Isaac, 2009, 2013), which cuts deep into Palestinian areas, are only some of the problems for Palestine directly resulting from its Israeli occupation.

The Security and Well-being of Palestinian Researchers

Issues of safety, security and stability are prominent in the narratives of Palestinian life and of researchers, particularly those in Palestine. Many Palestinian researchers see safety and security as the most important concerns they face when conducting research. In addition, the conflict occupies a central place in the lives of Palestinian male researchers. It is important to consider that these challenges are not limited to male researchers and that female researchers also face many challenges in conducting research in Palestine. However, the focus in this chapter is on Palestinian male researchers.

As noted by Kuttab (2006), and as expected, the quality of life and the living conditions of Palestinians have not improved; on the contrary, they have steadily deteriorated. Individuals, families and communities have experienced economic decline and threats to their personal safety and property through a set of policies designed to pauperise Palestinians and expand the Jewish presence on their land. For example, teachers' and students' lived daily reality is characterised by serious physical difficulties, traumatic events and humiliations (Giacaman *et al.*, 2002). For instance, I, as a male researcher, along with female researchers, have faced very serious physical challenges such as blockades, flying checkpoints and in some instances curfews that hinder us from travelling freely within the West Bank. According to Hammond (2007):

> the work permits of foreign visiting academics were tightly controlled, making it increasingly difficult for academics from Europe and America to work in Palestinian institutions. Palestinian academics (male and female) could not travel abroad and administrative arrests increased. Administrative detention became a regular aspect of campus life, as more students and teaching staff found themselves in detention without charge. (Hammond, 2007: 265)

Halper (2008: 152) uses the term 'Matrix of Control' to describe this state of affairs, a concept that is explicitly concerned with Palestinian immobility:

> The Matrix, an intricate and an interlocking series of control mechanisms, resembles the Japanese game of 'Go.' Instead of defeating your opponent as in chess, in 'Go' you win by immobilising the other side, by gaining control of key points of a matrix, so that every time your opponent moves he or she encounters another obstacle. (Halper, 2008: 152)

The Israeli policy of 'closures', another expression of collective punishments, was introduced a few months before the Oslo Accords were signed; however, it was systematically and steadily intensified during the years of the second *Intifada*. These closures were one of the main implications for me as a researcher, limiting my ability to conduct research and move around in various cities and towns in Palestine. Closure is intended to deny Palestinians their right to free movement, stemming from a 'pass system' first introduced in 1991, which required that every Palestinian had to obtain a colour-coded identification card and apply for a permit to move between and within what would eventually become Areas A–C (Tawil-Souri, 2011).

In the West Bank, Area A is under direct Palestinian control and includes the major populated cities but constitutes no more than 3% of the West Bank; Area B encompasses 450 Palestinian towns and villages, representing 27% of the West Bank, in jointly controlled territory in which the Palestinians can exercise civil authority but Israel retains security control; and Area C, in which Israel has exclusive control, constitutes the rest of the West Bank (70%), including agricultural land, the Jordan Valley, natural reserves, areas with lower population density, Israeli settlements and military areas (Hanafi, 2009, 2013). In the West Bank, closure is implemented through an agglomeration of policies, practices and physical impediments which have fragmented Palestine into ever smaller and more disconnected cantons (Weizman, 2007).

Hanafi (2009) argues that the Israeli colonial project is 'spacio-cidal' in that it designates land for the purpose of rendering an expected 'voluntary' transfer of the Palestinian population, mainly by targeting the space upon which the Palestinian people live. According to Hanafi, this policy constitutes two strategies: the 'space annihilation', to echo Kenneth Hewitt (1983), similar to that witnessed in Europe during WWII, such as the destruction of Dresden and Hiroshima; and ethnic cleansing, to use the words of Masalha (1992) and Pappé (2006), who both clearly illustrated how ethnic cleansing was not a circumstance of war, but rather a purported goal of combat for early Israeli military units directed by Prime Minister David Ben-Gurion, whom Pappé labels the 'architect of ethnic cleansing'.

Unlike researchers in most parts of the world, as a Palestinian male researcher I face various difficulties, which have been mentioned above. As a result, I mostly conduct research under extremely unusual circumstances and conditions. For example, problems with reaching study participant families for visits and the postponement of appointments are common difficulties faced during research (Abu Nahleh, 2006). Similar to all other Palestinians desiring to move throughout the region, researchers must adapt to the life left to them as a result of the Israeli military checkpoints and roadblocks, long-term and short-term curfews, invasions and assassinations, which are all enacted daily by the Israeli occupation. Space and time have acquired different meanings and associations, accordingly.

Access to higher education institutions has become very difficult for faculty because of long curfews and other restrictions on travel within

Palestine. In many cases, however, university faculty have begun to communicate with their students via the internet in an attempt to overcome Israeli-imposed restrictions on freedom of movement. In some cases, universities have tried to find alternative venues in neighbouring cities to hold classes and seminars. At the same time, most of the faculty at Birzeit University in Ramallah, West Bank, despite the fact that they sometimes do not received their full salaries for months on end, are also prepared to endure tremendous difficulties in order to meet with their classes. Simply attending school and holding classes have become major acts of defiance, turning universities and schools into sites of resistance. In addition, during the years I have been teaching at Bethlehem University, some of my master's students could not attend classes for a few days due to restrictions on movement, as they were coming from Hebron, a city on the West Bank.

Im/mobilites of Palestinian Male Researchers

Hammami's (2004) ethnography brilliantly describes the Surda checkpoint, which lies on the main road between Ramallah and Birzeit University:

> Commuters would disembark from the transit vans that jammed both ends of the no-drive zone. Skirting rubble and concrete blocks, they would trip down the valley and hold their breath as they passed Israeli soldiers, before finally trudging up the incline to the vans on the other side. Thousands made the walk every day. In the morning, the flow of fashionably dressed students on their way to the university crossed the flow of villagers heading into Ramallah for work and the services that can only be found in a city. In the afternoon, the pattern would repeat in reverse. Those who got thirsty along the way could grab a drink from a roving peddler. Others who had forgotten groceries could stop at one of the makeshift stands nicknamed 'the duty free'. (Hammami, 2004: 26–27)

Harker (2009) further describes the severities of these checkpoints:

> On the worst days, trigger-happy soldiers suddenly prohibited pedestrian traffic, and students and villagers were stranded on the wrong side of getting to work or home. More commonly, soldiers would drop in at the checkpoint for a few hours, to toy with the droves of walking commuters, stopping all – or a select few – for interminable identity card and baggage checks. (Harker, 2009: 19)

In my experience of travelling from Bethlehem to visit a site in Jericho or Ramallah, I have at times had to wait for two hours in my taxi at military checkpoints, roadblocks and closures. Travel time reflects new geographical realities lived by Palestinians in the 1967 occupied territories. I regularly travel to Palestine, around three times a year, to teach at Bethlehem University in Bethlehem and to conduct field research. Although I have dual nationalities, Palestinian as well as Dutch, particularly after the second *Intifada* uprising in 2002 I have had to travel through Jordan. This travel arrangement results in needing to spend at least one night in Amman, Jordan. The next morning, I depart from Amman at 7am, heading to the Jordanian-Israeli bridge-crossing border (Allenby Bridge), at which point I proceed through Jordanian checkpoints. Jordanian checkpoints are comparatively easy. After clearance, I board a bus to travel to the Israeli side. Sometimes we sit in the bus at the bridge for a few hours until the Israeli military give us the green light to move forward into the Israeli military checkpoint. Upon arrival at the Israeli checkpoint I must carry my own luggage to the scanning machine, before proceeding down a corridor for another body check and scan. After this, I go through a military identification desk which can take up to an hour. If I am lucky, it is 15 minutes before I can collect my passport and proceed through another checkpoint to pick up my luggage. Next, I board another bus to the Palestinian-controlled areas in the city of Jericho. Once in Jericho, I must again undergo a passport check, this time by the Palestinian Authority. Following this check, I can collect my suitcase and get into a taxi for the journey from Jericho to Bethlehem. On the road to Bethlehem (which used to be a 20-minute ride from Jericho), there are several more Israeli checkpoints.

As of January 2017, there were 88 checkpoints in the West Bank; these include 59 permanent checkpoints located deep within the West Bank, 18 of them in Area H2 in the city of Hebron where Israeli settlement enclaves have been established. Some of these checkpoints are constantly staffed, some only during the day or for part of the day and some are hardly ever staffed. Inspections at the checkpoints vary, but they are often random. Another 39 checkpoints are always staffed; these are in locations that are considered points of entry into Israel, although most are located several kilometres into the West Bank. Inspections at these checkpoints are rigorous. According to 2017 figures provided by the United Nations Office for the Coordination of Humanitarian Affairs (OCHA, as cited in B'Tselem, 2019), 2941 'flying' checkpoints were counted along West Bank roads (an average of 327 per month) from January until the end of September. A total of 5587 flying checkpoints were counted in 2016. These flying checkpoints control all Palestinian taxis travelling from Jericho to all Palestinian cities, including Bethlehem. At each flying checkpoint, the identity cards of all Palestinians travelling in the taxi, particularly those of men, are inspected, which takes an hour or more.

Feeling inferior and neglected are some of my emotions when travelling through occupied Palestine. The feelings of anger, but at the same time of despair, are overwhelming. The loss of masculine dominance, as well as the humiliation suffered at the hands of other (Israeli military occupation) men, generates a particular emotional geography for men of strong masculine disposition. The loss of masculine dominance at the multiple checkpoints is part of the emotion I (and likely others) experience as a Palestinian male researcher. All too common practices, such as being asked to remove clothing, sexual abuse, beatings or standing in handcuffs for hours under the sun, are some of the factors that have led to feelings of emasculation. In addition, such feelings of despair and frustration have been associated with 'abiding effects' (Clarke et al., 2006: 11), in this case, prolonged marginalisation and suffering in both male and female researchers as a result of the Israeli military occupation.

Many Palestinians, including me, living in and outside the West Bank, face serious obstacles to their mobility. According to OCHA, as of January 2017 there were 476 unstaffed physical obstacles along West Bank roads – including dirt mounds, concrete blocks and fenced-off segments. Of these, 124 were gates installed at the entrances to villages – 59 of them closed and 65 open most of the time, except when the military decides to close them (B'Tselem, 2019). This lack of mobility is a major challenge for those wanting to carry out research, as the decision to travel through or outside the West Bank, or when travelling from the West Bank (Bethlehem) to Europe or neighbouring countries such as Jordan (which I experienced in 2019; see also Barghouti, 2000; Halper, 2008) is not a simple task.

According to Abu Nahleh (2006), there is a common sense of fear among Palestinians that they will be detained or lose a child. This is in contrast to the portrayals by Israeli and Western media that sometimes show Palestinian mothers sending their children to be killed by the Israeli military occupation (Abu Nahleh, 2006: 115). There is an assumption by the Israeli occupation that a Palestinian male may be more dangerous than a female and, thus, male Palestinians have more to fear from the occupiers. Feelings of fear are an important element of life as a Palestinian male researcher. These emotions of fear are generated by unequal practices of power, as the relations between Palestinians and Israelis, where there is a colonised and coloniser, an occupied and occupier, are asymmetric. This marginalisation and exclusion is seen throughout Israel where embedded, systemic practices of racism preserve and elevate a Jewish majority, while demonising Arab and Palestinian minorities as second-class citizens (Isaac & Platenkamp, 2016). These minorities are said to be a 'threat' to the very existence of the majority way of life and culture. As Torabian and Miller (2017: 933) state, 'here, stereotypes and other racialised experiences speak to the control and unequal power relations between different groups, but germane to such perverse power relations is the understanding of the "Other"'.

In Said's (1978) seminal work, *Orientalism*, he tried to answer the question of why, when we think of the Middle East and Arabs, we have a preconceived notion of what kind of people live there, what they believe and how they act, even though we may never have been there or indeed met anyone from there. More generally, Said (1978) asked: How can we come to understand people, outsiders, who look different from us by virtue of the colour of their skin? The central argument of *Orientalism* is about the way in which we acquire this knowledge, and that the acquisition is not innocent or objective but, rather, highly motivated. Specifically, Said (1978) argues that the West (Europe and the American United States) look at the people of the Middle East through a lens that misrepresents the actual reality of those places and those people. Westerners create the lens, Orientalism, as a framework for understanding the unfamiliar and strange, making the people of the Middle East appear different and threatening.

As Moïsi (2009: 20) stated, '[in] the age of globalisation, the relationship with the Other has become more fundamental than ever', whereby Othering is now entangled within politics of fear and terror. Moïsi (2009) further contended that, since 9/11, the discourses of fear and a 'fear of the Other' have intensified in the West; non-Western countries, and more specifically members of Muslim and Arab communities, are now seen as the Others that the West should fear. Hooks (1992: 174) claimed: 'I think that one fantasy of whiteness is that the threatening Other is always a terrorist. This projection enables many white people to imagine there is no representation of whiteness as terror, as terrorizing.'

Checkpoint 300 Gilo

In this section I will describe the Gilo checkpoint in Bethlehem, and explain how gender inequality particularly affects those Palestinians who cross the border every day in order to reach their jobs in Israel. The characteristics of crossing the border for employment (work in Israel) are similar to those experienced by male Palestinian researchers and academics as they attempt to move about in the course of their field research in the occupied territories of Palestine. This context of encounters by the Palestinian workers in Israel will be used as a framework, because it has similarities to my experiences and difficulties as a male researcher conducting research in the Palestinian territories. Indeed, it is interesting to start with Palestinian workers crossing the Gilo checkpoint in Bethlehem, because it clearly illustrates the different obstacles and the emotions of insecurity, fear, inferiority and despair experienced by Palestinian workers in Israel, which reflect my own lived reality as a Palestinian male researcher.

The checkpoint is built into the Palestinian West Bank Segregation Wall and can be entered through three lanes: one lane for people who have 24-hour permits; a lane for Palestinian workers with eight-hour permits; and a humanitarian lane, which is intended for women, children and elderly Palestinians. This checkpoint in Bethlehem is one of the most intensively crossed checkpoints in occupied Palestine, used mainly by Palestinians heading from south of the West Bank on their way to Jerusalem and Israel. Checkpoint 300 was categorised as a 'terminal checkpoint' by the Israeli occupation in 2005 (Rijke & Minca, 2018). In November 2005, the Israeli military inaugurated terminal Gilo 300 at the northern entrance of Bethlehem. Movement between Jerusalem and Bethlehem is supposed to take place through this terminal.

Inside the terminal, people are classified according to the ID documents they carry. Passing from Bethlehem to Jerusalem entails passing through several checkpoints where Israeli military communicate with the public through glass with the use of microphones to check their documentation. Checkpoint 300 is an example of a new generation of installations described by Mansbach (2009) as an attempt to 'demilitarise' the checkpoints and normalise the Israeli military control of the mobility of Palestinians. Checkpoint practices, from the perspective of the Palestinians passing through Checkpoint 300 in Bethlehem, contain very important characteristics that are similar to those I have described facing as a Palestinian male researcher. These checkpoints are a means of surveillance, selectively limiting the mobility of Palestinians, but they are also used by the Israeli military to ensure that the capacity of Palestinian people to reach their daily destination is never entirely predictable (Rijke & Minca, 2018). For Palestinian men, everyday life begins in the crowd that forms as early as 3am when hundreds of workers push for position before the gate opens to the eight-hour permit lane at 5am (Griffiths & Repo, 2018). A 2009 report in the *Belfast Telegraph* (cited in Griffiths & Repo, 2018: 21) states, 'before sunrise ... there is scuffling when the tempers of the men, many of whom have been up since 3am, begin to fray as they compete to squeeze into the alley to queue for a lengthy series of Israeli security checks of their IDs, work permits and biometric palm prints'.

Generally, it takes 10–20 minutes before Palestinian males are able to progress through these bottlenecks at the entrance and file into the cage-like structure that has been created as part of the Gilo checkpoint, now called a 'terminal'. As Griffiths and Repo (2018) report,

> tension and frustration are inevitable in this environment that foments – even encourages – scuffles and the fraying of tempers. Once inside the entrance, the scuffles tend to dissipate as positions in the queue become more or less fixed and the crowd moves in a stop-start fashion. Thus begins the daily commute that compels Palestinian men to leave the home each day as early as 2am and return before 7pm on the expiration of their 8-h permits (or face a night in a prison cell). (Griffiths & Repo, 2018: 21)

Such practices associated with checkpoints act as spatial-temporal devices for imposing on the daily lives of Palestinian men. As Weizman (2007: 151) has noted of checkpoints across Palestine: 'soldiers regulate the pace of passage by using an electrical device that controls the turning of gates ... [e]very few seconds soldiers stop the rotation of the turnstiles, so that several Palestinian men remained caged between the gates.' Similar to the process I have already described, all Palestinian men depend on a minibus link waiting for them at the other side of the Gilo terminal to reach their ultimate destination. Delays may result in a missed day of work and/or loss of pay, as bus drivers do not accommodate late arrivals (B'Tselem, 2016).

Men may be deprived of eligibility for an eight-hour working permit for a number of reasons, such as: having been in an Israeli prison; having been involved in political party, faction or movement activities; being under the age of 30 or over 50; or being unmarried and without at least one child. All Palestinian men who meet these criteria are 'blacklisted' on their ID cards and are not allowed to enter Israel or even travel between different cities on the West Bank; sometimes they are also not permitted to travel abroad (Griffiths & Repo, 2018). Thus, if you are an activist or have another blacklisted characteristic and are an academic conducting field research in Palestine, obtaining a working permit can be nightmarish. Because of the potential for interrogation and humiliation, these checkpoints have been described as 'an assault on the flesh' (Aaltola, 2005: 270). The hours I spend at checkpoints when visiting Palestine for the purpose of conducting research are consequently characterised by depressive emotions that can render me too tired to work – unable to think, much less able to organise or consider more important perspectives.

Conclusion

In this chapter I have described how, as a Palestinian male researcher, I face numerous difficulties when conducting research in the occupied territories of Palestine. Drawing on my personal account and fieldwork experience, particularly in the sociopolitical context of Palestine, this chapter has exposed my own experiences as a male researcher and how these have influenced my personal safety, security, access, movement restrictions and well-being. In addition, this chapter addresses more open-ended questions around 'doing research' for a male Palestinian in a conflict-ridden destination, and the potential threats faced by a male Palestinian researcher. For me as a male Palestinian researcher, passing through an Israeli checkpoint is a daily ritual I cannot avoid on my way to a university or in conducting field research in Palestine.

Nevertheless, the checkpoints are key sites where the impact of the occupation is felt on a daily basis. Despite the limited literature on this relevant topic, I have attempted to analyse my daily experience as a male researcher by focusing on the immobilities and insecurities of male Palestinians at checkpoints. I have sought to examine the impediments regulating Palestinian men (and women) because they have characteristics similar to those faced by male researchers in Palestine – both within and beyond the security apparatus. The checkpoints and obstacles facing Palestinian people play an important role in Palestinian society, including in the lives of male researchers. For men doing field research in the Palestinian territories, the waiting hours at checkpoints and closures mean that community and family life is interrupted and curtailed. Male researchers travelling within Palestine could be arrested or even killed at certain checkpoints. Being attacked by Israelis who have settled near Palestinian villages and towns is becoming commonplace. Israeli settlers are nowadays guarded by Israeli military occupation. Therefore, my family, and others, are always under pressure about what might happen during travel within the West Bank.

The checkpoints and blockades work to disrupt social relations within family life and Palestinian society as a whole. The long hours of commuting required by checkpoints and closures means that as husbands and fathers we are left with little time to spend with our wives, children, friends and wider family. Previously when I travelled to Palestine, a simple visit to a family member took around 20 minutes, but nowadays as a result of the Israeli Segregation Wall, the same visit takes at least 50 minutes; on top of this delay, passage may ultimately be denied. The Wall separates Palestinians from each other.

Further inquiry will build on these findings. Within this contribution, I hope to encourage further research on male researchers doing field research in Palestine and beyond in order to examine how border closures and lack of mobility impact their everyday life routines.

References

Aaltola, M. (2005) The international airport: The hub-and-spoke pedagogy of the American empire. *Global Networks* 5 (3), 261–278.

Abu Nahleh, L. (2006) Six families: Survival and mobility in times of crisis. In L. Taraki (ed.) *Living Palestine: Family Survival, Resistance and Mobility under Occupation* (pp. 103–184). New York: Syracuse University Press.

Barghouti, M. (2000) *I Saw Ramallah*. New York: Anchor Books.

Bianchi, R.V. and Stephenson, M.L. (2013) Deciphering tourism and citizenship in a globalized world. *Tourism Management* 39, 10–20.

Bondi, L. (2005) Making connections and thinking through emotions: Between geography and psychotherapy. *Transactions of the Institute of British Geographers* 30, 433–448.

Brownlow, A. (2005) A geography of men's fear. *Geoforum* 36, 581–592.

B'Tselem (2016) *Inhuman Conditions in Checkpoint 300*. Jerusalem: B'Tselem – The *Israeli Information Centre for Human Rights* in the Occupied Territories. See https:// www.bstelm.org/video/20170731_inhuman_conditions_in_checkpoint_300 (accessed 1 January 2019).

B'Tselem (2019) *Restrictions on Movement*. Jerusalem: B'Tselem – The *Israeli Information Centre for Human Rights* in the Occupied Territories. https://www.btselem.org/video/20160731_inhuman_conditions_in_checkpoints_300 (accessed 12 February 2019).

Buda, D. (2012) Hospitality, peace and conflict: Doing fieldwork in Palestine. *Journal of Tourism and Peace Research* 2 (2), 50–61.

Clarke, S., Hoggett, P. and Thompson, S. (2006) *Emotion, Politics and Society*. Basingstoke: Palgrave MacMillan.

Cole, S. and Eriksson, J. (2010) Tourism and human rights. In S. Cole and N. Morgan (eds) *Tourism and Inequalities: Problems and Prospects* (pp. 107–125). Wallingford: CABI.

Connell, R.W. (2005) *Masculinities* (2nd edn). Cambridge: Polity Press.
Crang, M. (2003) Qualitative methods: Touchy, feely, look-see? *Progress in Human Geography* 27 (4), 494–504.
Davidson, J., Bondi, L. and Smith, M. (2005) *Emotional Geographies*. Aldershot: Ashgate.
Day, K. (2006) Being feared: Masculinity and race in public space. *Environment and Planning A* 38, 569–586.
Giacaman, R., Abdullah, A., Abu Safieh, R. and Shamieh, L. (2002) *Schooling at Gunpoint: Palestinian Children's Learning Environment in Warlike Conditions – The Ramallah/al Bierh Urban Centre*. Birzeit: Birzeit University, Institute of Community and Public Health.
Griffiths, M. and Repo, J. (2018) Biopolitics and Checkpoint 300 in occupied Palestine: Bodies, affect and discipline. *Political Geography* 65, 17–25.
Hall, C.M. (2010) Equal access for all? Regulative mechanisms, inequality and tourism mobility. In S. Cole and N. Morgan (eds) *Tourism and Inequality: Problems and Prospects* (pp. 34–49). London: CABI.
Hall, D. and Brown, F. (2010) Tourism and welfare: Ethics, responsibility and wellbeing. In S. Cole and N. Morgan (eds) *Tourism and Inequality: Problems and Prospects* (pp. 143–161). Wallingford: CABI.
Halper, J. (2008) *An Israeli in Palestine: Resisting Dispossession, Redeeming Israel*. London: Pluto Press.
Hammami, R. (2004) On the importance of thugs: The moral economy of a checkpoint. *Middle East Report* 231, 26–34.
Hammond, K. (2007) Palestinian universities and the Israeli occupation. *Policy Futures in Education* 5 (2), 264–270.
Hanafi, S. (2009) Spacio-cide: Colonial politics, invisibility and rezoning in Palestinian territory. *Contemporary Arab Affairs* 2 (1), 106–121.
Hanafi, S. (2013) Explaining the spacio-cide in the Palestinian territory: Colonization, separation, and state of exception. *Current Sociology* 61 (2), 190–205.
Harker, C. (2009) Student im/mobility in Birzeit, Palestine. *Mobilities* 4 (1), 11–35.
Hewitt, K. (1983) Place annihilation: Area bombing and the fate of urban places. *Annals of the Association of American Geographers* 73 (2), 257–284.
Higgins-Desbiolles, F. (2006) More than an 'industry': The forgotten power of tourism as a social force. *Tourism Management* 27 (5), 1192–1208.
Hooks, B. (1992) *Black Looks: Race and Representation*. Boston, MA: South End Press.
Isaac, R.K. (2009) Alternative tourism: Can the segregation wall be a tourist attraction? *Tourism and Hospitality, Planning and Development* 6 (3), 247–236.
Isaac, R.K. (2013) Palestine: Tourism under occupation – the ramifications of tourism in Palestine. In R. Butler and W. Suntikul (eds) *War and Tourism: A Complex Relationship* (pp. 143–158). London: Routledge.
Isaac, R.K. and Platenkamp, V. (2016) Concrete U(dys)topia in Bethlehem: A city of two tales. *Journal of Tourism and Cultural Change* 14 (2), 150–166.
Isaac, R.K., Hall, C.M. and Higgins-Desbiolles, F. (2016) *The Politics and Power of Tourism in Palestine*. Abington: Routledge.
Jamal, T. and Hollinshead, K. (2001) Tourism and the forbidden zone: The underserved power of qualitative inquiry. *Tourism Management* 22 (1), 63–82.
Johnston, L. (2005) *Queering Tourism: Paradoxical Performances at Gay Pride Parades*. London: Routledge.
Johnston, L. (2007) Mobilizing pride/shame: Lesbians, tourism and parades. *Social and Cultural Geography* 8 (1), 29–45.
Kuttab, E. (2006) The paradox of women's work: Coping, crisis and family survival. In L. Tarakai (ed.) *Living Palestine: Family Survival, Resistance and Mobility under Occupation* (pp. 231–276). New York: Syracuse University Press.
Lyon, D. (2003) *Surveillance after September 11*. Cambridge: Polity Press.

Mansbach, D. (2009) Normalising violence: From military checkpoints to 'terminals' in occupied territories. *Journal of Power* 2 (2), 255–273.

Mansvelt, J. and Berg, L.D. (2005) Writing qualitative geographies, constructing geographical knowledges. In I. Hay (ed.) *Qualitative Research Methods in Human Geography* (pp. 248–265). Oxford: Oxford University Press.

Masalha, N. (1992) *Expulsion of the Palestinians: The concept of 'transfer' in Zionist political thought, 1882–1984*. Washington, DC: Institute for Palestine Studies.

McIntosh, A.J. (2010) Situating the Self in religious tourism research: An author's reflexive perspective. *Tourism: An International Interdisciplinary Journal* 58 (3), 213–227.

Meth, P. (2009) Marginalised men's emotions: Politics and place. *Geoforum* 40, 853–863.

Meth, P. and Bondi, L. (1999) Embodied discourse: On gender and fear of violence. *Gender, Place & Culture* 6 (1), 67–84.

Moïsi, D. (2009) *The Geopolitics of Emotion: How Cultures of Fear, Humiliation, and Hope are Reshaping the World*. New York: Doubleday.

Mura, P. (2010) 'Scary ... but I like it! Young tourists' perceptions of fear on holiday. *Journal of Tourism and Cultural Change* 8 (1), 30–49.

Pain, R.H. (2001) Gender, race, age and the fear in the city. *Urban Studies* 38, 899–913.

Pain, R.H. (2003) Youth, age and the representations of fear. *Capital and Class* 80, 151–171.

Pappé, I. (2006) *The Ethnic Cleansing of Palestine*. Oxford: Oneworld.

Rijke, A. and Minca, C. (2018) Checkpoint 300: Precarious checkpoint geographies and rights/rites of passage in the occupied Palestinian territories. *Political Geography* 65, 35–45.

Said, E. (1978) *Orientalism*. New York: Pantheon Books.

Sharp, J. (2009) Geography and gender: What belongs to feminist geography? Emotion, power and change. *Progress in Human Geography* 33 (1), 74–80.

Shirlow, P. and Pain, R. (2003) The geographies and politics of fear. *Capital and Class* 80, 15–26.

Tawil-Souri, H. (2011) Qalandia checkpoint as space and non-place. *Space and Culture* 14 (1), 4–26.

Torabian, P. and Miller, M.C. (2017) Freedom of movement for all? Unpacking racialized travel experiences. *Current Issues in Tourism* 20 (9), 931–945.

Tribe, J. and Liburd, J. (2016) The tourism knowledge system. *Annals of Tourism Research* 46, 245–255.

Tucker, H. (2009) Recognising emotions and its postcolonial potentialities: Discomfort and shame in a tourism encounter in Turkey. *Tourism Geographies* 11 (4), 444–461.

Waitt, G., Figuero, R. and McGee, L. (2007) Fissures in the rock: Rethinking pride and shame in the moral terrains of Uluru. *Transactions of the Institute of British Geographers* 32 (2), 248–263.

Weizman, E. (2007) *Hollow Land: Israel's Architecture of Occupation*. London: Verso.

Part 4
Paternal Masculinities

12 Finding Gender at the Intersection of Family and Field: Family Presences in Sweden

Stuart Reid

Introduction

In the 1980s, feminist scholars alerted us to the artifices of gender by 'making femininity and masculinity problematic' (Kimmel, 1993: 30). Initially, masculinity was mostly seen as a structural condition of unequal relations among women and men, exemplified by the 'invisible masculinity' of Kimmel (1993) and the 'hegemonic masculinity' of Connell (2005 [1995]). In Kimmel (1993), 'invisible masculinity' stood for the reproduction of inequalities stemming from normalisation of the white heterosexual male as the putative gender standard, the privileged position rendered invisible to, and thus reproduced by, men who 'have come to think of themselves as genderless' (Kimmel, 1993: 30). In Connell (2005 [1995]), 'hegemonic masculinity' described problematic relations of dominance among men and women resulting from making certain performances of maleness seem natural and normal. In pointing to the normalisation of normative gender arrangements, these perspectives indicated social practices as the fundamental ingredient of gender arrangements. Here, West and Zimmerman (1987) saw gender as an undertaking of women and men who were 'doing gender' by 'managing situated conduct, in the light of normative conceptions of attitudes and activities appropriate for one's sex category' (West & Zimmerman, 1987: 127). From this performative vantage, gender was 'not something that people are, but something that they do' (Berglund *et al.*, 2018: 3), as much a social 'face' as any other (Goffman, 1956), a social construction (Butler, 2004). Thus, there is not one masculinity but many masculinities (Connell, 2005 [1995]), and they are socially contingent. What this means for me, as a white, Western middle-aged man, is that my embodied persona does not exemplify some typical masculinity; instead, my male-sexed human vessel stands as a purveyor of assorted potential masculinities, their activation circumstantially contingent.

However, knowing of these potential masculinities and finding them are two different things. The complexities of gender make for a difficult reflexive task (Porter & Schänzel, 2018), as masculinities (and femininities) not only come in different shapes and sizes in 'the field' but are entangled with other positions, all constantly (re)negotiated in the field (Sultana, 2007; Swanson, 2018). Circumstantially called to life, our multiplex gender positions are thus slippery to grasp, appearing as ephemeral will-o'-the-wisps variously active in and out of the field. As a heterosexual married man who is a father to three young children I occupy family roles as 'husband' and 'father', and these circumstantial positions variously activate masculinities of father-hood and husband-hood in all the domains of my social life, including when I am a 'researcher' in the field.

Although reflexive accounts have challenged the myth of the lone, male ethnographer (e.g. Cornet, 2013; Frohlick, 2002; Korpela *et al.*, 2016), the personal and professional entanglements of family and field have largely escaped attention (Korpela *et al.*, 2016), 'many ethnographies being written as if the ethnographer had no ties and could disappear for long periods of time in faraway villages to undertake research' (Cornet, 2013: 80). When such family-field entanglements have been discussed, it has mainly been from feminine vantages (e.g. Canosa, 2018; Cornet, 2013; Farrelly *et al.*, 2014; Frohlick, 2002; Khoo-Lattimore, 2018; Korpela *et al.*, 2016; Levey, 2009; Porter, 2018; Swanson, 2018), and while there is much to be gained from these feminine accounts, the masculine vantage is lacking (Porter & Schänzel, 2018). Furthermore, although the spatial boundaries of such entanglements are sometimes mentioned in reflexive accounts (e.g. Frohlick, 2002), it seems that the spatial dimensions of these epistemological entanglements have not been fully addressed: for instance, while some accounts offer insights into the epistemological implications of family co-presence *in* the field (e.g. Frohlick, 2002; Korpela *et al.*, 2016), fewer extend to contemplation of symbolic family presences such as wedding rings (Swanson, 2018) or pregnancy (Porter, 2018). As outward manifestations of familial presence, such 'absent presences' point to the tricky matter of when and where the entanglements of family and field might arise. Lastly, it seems to me that while reflexive accounts have tended to offer insight into the influence of family in the field, less has been said about the field in the family. Therefore, in contemplating my reflexive account of my masculine entanglements at the intersection of family and field, I have been provoked to wonder *where* I might draw the lines demarcating the masculinities of field-family entanglements. As gender is part and parcel of every aspect of our social lives, it seems to me that the influence of my family does not start and end with my being *in* the field, but extends to 'outside' and 'before' and 'after', raising the broader challenge of accounting for the gender positions of family in my relation *to* the field (Swanson, 2018). Thus, reflexivity calls me to account for masculinities in both my professional and personal domains, looking for the masculine entanglements of family-in-field and field-in-family. In undertaking my complicated reflexive task, I draw inspiration from Farrelly *et al.* (2014) who, in pointing to the ephemeral constitution of motherhood as the co-presence of 'being there', indicate a helpful thematic of absence-presence. So I will lean on this thematic to help me excavate the traces of gender in the masculinities arising at the intersection of field and family. In the spirit of honest knowledge (Haraway, 1988; Swanson, 2018), I offer my reflections on these gender influences as a white, heterosexual man who is a married father of three young children. Working from this vantage, I seek to offer a partial response to, and support for, the call for 'true gender research' that stems from 'recognition of the intersection between the gender roles of mothers and fathers, and their femininities and masculinities' (Schänzel & Smith, 2011: 144).

My reflections draw from my experiences as a doctoral student investigating a phenomenon labelled 'lifestyle entrepreneurship'. As is the case for any other socially constructed artefact of language, this label carries various significations, chief among these being that 'lifestyle entrepreneurs' are thought to accord particular weight to various personal concerns in their enterprise of business (Carlsen *et al.*, 2008; Morrison, 2006). Ostensibly striving for desired 'life quality' (Marcketti *et al.*, 2006), these enterprisers set profit-making in the context of wider life aims, often making just enough money to 'get by' (Ateljevic & Doorne, 2000). Although lifestyle enterprising has been noticed in contexts outside tourism (e.g. Burns, 2001; Cederholm, 2015; de Wit Sandström, 2018; Marcketti *et al.*, 2006), it is said to be prevalent among small enterprises in tourism (Peters *et al.*, 2009), at least since Shaw and Williams (1987) identified it in their study of small seaside hotels in Cornwall. Suffice to say, conventional wisdom holds that small enterprises serving tourists are often sites of lifestyle enterprising. With recent sociological perspectives affording fresh insights into lifestyle enterprising as a sociocultural phenomenon (see, for example, Ateljevic & Doorne, 2000; Cederholm, 2015; Cederholm & Åkerström, 2016; Cederholm & Hultman, 2010; Hultman & Cederholm, 2010), I took this sociological perspective as the point of departure for my field engagements with lifestyle enterprise in connection with my doctoral studies.

Family in the Field

Masculinities in formation

With the abundance of lifestyle enterprising rendering the world 'my oyster',[1] I was able to entertain my desire to conduct fieldwork somewhere other than in my home in Australia. That the familiarity of local cultural norms might blind me to important nuances in the sociocultural construction of lifestyle enterprising was only part of the explanation; the other was simply my desire to experience the challenge of doing my fieldwork 'someplace else'. Perhaps this desire stemmed from some masculine urge to prove my professional worth by 'conquering' the unknown. Regardless of the cause, I shortly found my professional urge for exoticism conflicting with my private urge to be present in my family. That is, I simply could not countenance lengthy absences from them – a sentiment echoed by other researchers, albeit from feminine vantages (e.g. Frohlick, 2002; Korpela et al., 2016; Porter, 2018). In my mind, absence from my family was a double negative: not only would it have deprived *me* of *their* presence, but it would have also undermined my perceived masculine duty to 'be present' in my family (Farrelly et al., 2014). So I mulled over the possibility of solving this personal–professional dilemma by bringing my family with me to the field.

I was aware that relocating a family of five, including three young children, was a monumental logistical feat necessitating contemplation of many interests besides my selfish impulse to do my fieldwork somewhere else. Chief among my concerns was my perceived masculine duty (as father and husband) to keep my family happy and safe. The solution to this dilemma came serendipitously, in the form of an opportunity to undertake my doctoral studies at Lund University in the famously family-friendly nation of Sweden, and with that my exotic field resolved. So even *before* my research had begun, masculinities of fatherhood and husbandhood had stamped their marks upon it, delimiting the vast scope of the potential (global) field and setting the stage for 'accompanied fieldwork' (Frohlick, 2002) in Skåne in southern Sweden.

Introducing the field

Skåne is the southernmost region of Sweden (Figure 12.1). In the southeastern corner lies an area known as Söderslätt (literally 'South Plain'), its fertile soils supporting abundant farming enterprises and rendering multihued vistas 'pretty as a patchwork quilt' (Tourism Skåne, 2019b). In the southwest lies Österlen, an area 'synonymous with art, sky, and sea' (Tourism Skåne, 2019c). Possessed of picturesque coastal villages and reputedly the best beaches in Sweden (Sandhamnen), it is a haven for artists and *bons vivants*. With 'fertile farmlands, lush forests and clear blue lakes … inviting sandy beaches … rural areas and the charming fishing villages' (Tourism Skåne, 2019a) and linked to Denmark by the Oresund bridge, southern Skåne is a popular tourist and lifestyle destination, its small enterprises offering great promise to encounter the lifestyle enterprising of empirical interest to me.

Masculinities in relocation

In packing our lives into little suitcases and getting on a plane (Figure 12.2), we effectively said goodbye to the familiar realm of our

Figure 12.1 Map of Southern Skåne
Source: Original map provided by Tourism Skåne.

home; in relocating to Skåne, each of us suffered from the culture shock that went with being in the field (DeWalt & DeWalt, 2010). However, the dislocations of culture shock differed. Here, the luxury of occupational purpose afforded me the privilege of viewing the experience not solely from my personal perspective, but from the professional vantage of my researcher-hood; for me, and me alone, the new place was a 'dual-purpose site' (Di Domenico & Lynch, 2007) – both a place of 'work' in the field and (eventually) a place of 'home'. So although our foreignness was rendered apparent to us all in the daily exigencies of life in a new place (Wylie,

Figure 12.2 Departing Brisbane, Australia in 2017 with suitcases in tow

1987), I alone enjoyed the privilege of respite in professional purpose. As Wylie (1987) suggests, accompanying family members are unlikely to see the field as anything other than being 'somewhere else' and I would say that this was true for us. As my family did not enjoy my occupational privilege, they encountered the challenges of relocation solely as foreigners acclimating to a new place. I would suggest that relocation and dislocation are close bedfellows and that accompanying family members experience relocation to the field wholly as spatial-temporal dislocation from 'home'.

On occasion (and particularly in the beginning), certain events would transpire that made our foreignness vastly more apparent to us; these events rendered experiences of dislocation acute and inspired longings for the absent familiarity of our former home. Although we all experienced this, adults and children alike, it was the children who seemingly suffered most, for in relocating to my field they suffered the double injustice of enforced and lengthy dislocation – not only was their sentence of *absence* from home inflicted upon them by the decisions of us as parents (Wylie, 1987), and caused by me, but their relative youth made it seem a relatively long one (while the few years of our planned relocation did not seem long to us adults, it seemed interminably long to the children). Furthermore, the dislocation effects of relocation were not uniform among our children, seemingly striking our eldest daughter (eight-year-old Alice) more than her younger twin siblings (four-year-old George and Megan). Although individual temperaments were at play, the violence of dislocation may have been generally rendered differently according to their ages. Wylie (1987) suggests that for younger children the world revolves more closely around their parents, so perhaps the comforting presence of family dulled the dislocation experience for Megan and George relative to their older sister Alice, for whom the social world was wider. Their varied expressions of distress brought the blurred terrain of my professional and personal domains to the fore. As my professional domain was the cause of the dislocation malaise afflicting my children, their distress evidenced harms inflicted upon them by me, challenging my masculine ideals of fatherhood. So the personal arena of parent-child relationships is another space where masculinities can be activated in the field.

Although I did my bit to help with the various demands of our relocation, it was my wife Sam who did the 'heavy lifting' in facing the practical challenges of life in the exotic place that was my field. So while it was my professional domain that imposed the dislocation of relocation, it was her work that defrayed it by reconstructing 'home'. Although this division was ostensibly the result of my need to engage in the work that brought me to the field, my masculine sensibilities of husband-fatherhood were entangled in this too: in doing my work I was not only fulfilling my professional *raison d'être* for being in the field, but also my perceived duty (as husband and father) to earn an income and provide for the material needs of my family. So, with the cultural totality of economically dominated modern life entwined with my personal and professional duties, I went off to work and it was left to my wife to do the work of family care and home-making that went with our relocation. In enabling my work *and* supporting our family, her work upheld the institutions of both work and family. I would suggest that the work of homemaking and family care in the field could be seen as a kind of 'relocation work' that supports the researcher in their professional and private spheres by upholding both work and family.

Others have commented on the gendering of care work in the context of family relocations of various durations, such burdens being typically borne by women wherever they may be located, be it when embarked upon the temporary relocations of 'vacation' (Schänzel & Smith, 2011; Small, 2005) or the more permanent relocations of the field (Wylie, 1987: 108). In my experience, this gendered burden existed in the foreign place that was my field, where for one reason or another, my wife ended up carrying the burdens of the 'field relocation work' caused by (and benefiting) me. This leads me to suggest that the family care and homemaking burdens of field relocation work are both uneven and gendered, privileging the one who is the researcher at the expense of the one who is not, and likely all the more so when the researcher happens to be male.

Although the assorted dislocation effects of our family relocation diminished with the passage of time, as the exotic place I called 'the field' gradually became less foreign and more like 'home' (largely thanks to the field relocation work performed by my wife), the intermediate expressions of familial distress were personally challenging for me. Not only did each instance of family distress render the culture shock cost of accompanied fieldwork apparent (DeWalt & DeWalt, 2010), but in my mind each also signalled a personal failure to acquit my masculine duties (as father and husband) to protect my family from harm. In troubled moments I confronted the entanglement of my personal and professional domains, often reweighing the wisdom of subjecting my family to the stresses that came with *my* exotic fieldwork project. My experiences in this suggest to me that the dislocation stresses of the field are not confined to the institution of work, but variously spill into the personal domain to tug at the fabric of the family. While working through these challenges did strengthen our family institution, bringing us closer together as we variously leaned on each other for support, I would join with Wylie (1987) to caution that this might not always be so – the tidal forces of relocation-dislocation that tug at the fabric of the family institution may also tear it apart.

Masculinities in selection

Attending to my roles as 'researcher' and 'husband-father', I effectively chose to limit my fieldwork to areas near our new home-away-from-home so I would not be absent from my family. I reconciled the choice professionally by reasoning that nothing would be lost to the research by doing so, the field sites in southern Skåne being as good as any. Whether the choice materially affected my research or not is a moot question, for as with any choice in life, the road not travelled remains unknown and one can only speculate as to what might have happened had different choices been made. Yet it remains that my choice was gendered, my masculine ideals at play in the choices limiting the geographic scope of my sites in the field. Although it is common practice to justify one's choices of fieldwork sites on rational criteria linked to scientific objectivity, such as accessibility, time or cost, I would suggest that honest reflection will soon relegate such claims to the status of window dressing. If researchers are honest about such things, they will admit that various personal considerations shape all such choices, and gender is always among them.

Masculinities in apprehension

As the lifestyle enterprising phenomenon is abundant in tourism settings, I reasoned that small enterprises offering services to tourists in southern Skåne were good places to find it. However, since most of these were part-time enterprises, often open during weekends, I confronted my masculine ideals in the dilemma of absence from family during the time typically reserved for family. As these enterprises were open to the public, a family presence was to be expected, so I reconciled my dilemma by merging the roles of researcher and husband-father and bringing my family with me in many of my visits to the field (Figures 12.3 and 12.4).

The resulting juxtaposition of my personal and professional domains revealed, among other things, the mutability of both place meanings and

Figure 12.3 My family interacting with one of my participants in the field

Figure 12.4 Family in my field – assorted field sites in southern Sweden

roles in familial accompaniment. Family co-presence brought my masculinities into play, rendering the meanings of roles and places malleable, the constitution of each essentially rendered in dialectic and diachronic unfoldings of presence and purpose. Specifically, as researcher I was 'in the field', but when I was engaging with my wife and children at these places I was as much a tourist as any other, variously 'husband' and 'father' as circumstantially called into being – the periodic demands of 'Daddy look at me!', 'Dad look at this!' 'Daddy can I have …' reconstituting me from 'researcher' to 'father', reconfiguring place from field-site to tourist-site in the process. The upshot in terms of my research was that the presence of family changed who I was and what I could and did see in the field. My fieldwork notes and interviews are replete with traces of such family influence.

A typical example is found in a visit to a small enterprise in the countryside one weekend. As it was a Sunday, I suggested we should all go – a suggestion allowing me to offset the guilt of working at the weekend (a time usually reserved for family). Upon our arrival, we surveyed a cluster of buildings in a rural setting, one of which was signed '*butik*' (shop), and so we all went in there. Inside, a young man stood at a glass counter and an older man and woman were working in the background, with the counter serving as a kind of boundary dividing the 'frontstage' of the customer area from the pseudo 'backstage' where a less public work was done (Goffman, 1956). Greeting us, the young man invited us to taste some samples, and during this interaction I asked if he made the products we were tasting, to which he replied 'no, she does', indicating the woman, his statement prompting her to respond, 'yes, I am the *mom* and I make it', before smiling and adding that he was 'a good *son* for helping out today!' At the time, I wondered if this detail would have been shared had

I not been seen, thanks to the presence of my children, as a fellow-parent – my 'fatherhood' affording a common bridge of parenthood and thus furnishing insight into the familial entanglements of her enterprising. Another example is found in another event at that same enterprise when, while playing in the garden, one of my children declared an urgent need to use the toilet. As no toilet was visible, I re-entered the shop to ask for help in locating one. In response to my query, the woman beckoned me to join her in the pseudo backstage (Goffman, 1956) behind the counter, whereupon she pointed through the window to a door in another building and told me I could find a toilet in there. Upon passing through the indicated door, I realised I was in the private space of her home, and I was struck by the short distance between home and business. By activating my fatherhood, my child had secured my access to that private realm, affording me insight into the entanglements of 'home' and 'business' in the woman's enterprising. In these and many other instances, my wife and children were incidental adjuncts to my research enterprise, all 'little wedges' (Wylie, 1987) opening spaces for me in the field by variously activating masculinities of father-hood and husband-hood.

Family influences happened in other ways too. When accompanying me on other site visits and 'go alongs' (Kusenbach, 2003), my children would often engage with the people in my field, and by noticing things that only children see and asking questions only children ask, they drew my attention to things I might have otherwise missed. My wife also interacted with the people in these places, seeing different things and asking different questions, including those I might not have thought to ask, or perhaps could not have asked in the same way, as a man. Through their active involvement in these places they were all accidental 'adjuncts' (Wylie, 1987) to my research enterprise; their in-field engagements reconfigured them into de facto research assistants, affording me different insights into the spaces of my field.

The physical presence of family members at my field sites was not always smooth sailing. Our children were sometimes difficult to handle, their assorted behaviours and demands variously challenging our capacity as parents in the private domain and affecting my fieldwork in the professional domain. However, these disruptions and disturbances often yielded valuable insights. One example is found in an event transpiring one weekend when my momentarily restless son trod on an electronic door chime positioned on the floor near the door of a shop. Finding myself suddenly recast as a 'father', I grumbled at my son to 'quiet down' and offered my apology to the unruffled enterpriser, who assured me it was no bother as she was well acquainted with the antics of small children, both as mother herself and in her other job as a primary school teacher when she wasn't doing *this*. Thus, my son's seeming 'disruption' afforded insights into the entanglements of her personal and professional roles in relation to her part-time enterprising. Another good example occurred one day when our children, rushing outside to play, inadvertently locked us into an orangery (in Sweden this is a glass house providing a sheltered spot to sit in a garden), this event causing the husband of the female enterpriser to 'rescue us' by opening the door from the outside. Making space for him to stop and chat with us, the ensuing discussion afforded me many insights into the familial entanglements of their enterprising. My fieldwork notes and recordings hold many examples of insights emerging from such seemingly 'disruptive' events. There is much knowledge to be gained from disruptions and failures in the field (Levey, 2009), and this includes the happenstance activation of gender positions stemming from family co-presence. My experience challenges the assumption of family as a negative force in the researcher's field, the 'disruptions' of family having afforded me many, sometimes remarkable, insights into my field. In this respect, it bears mentioning that the work of research is essentially processual – the unfolding engagements with the field do not reveal their secrets on cue, but haphazardly and happenstance – and in my experience serendipitous insights arise not when and where we plan them, as if we could script them, but when and where they will. So when it comes to family co-presence in the field, I would suggest a more productive posture that the distractions and disturbances that go with family accompaniment are neither good nor bad, but merely ingredients of the unfolding scene, fomenting different events and rendering different insights. The temporality of fieldwork with family also merits mention here; although fieldwork with children *can* take longer (Levey, 2009), family presences can sometimes speed things up.

I did not always take my family with me in the field, but their lack of physical presence did not remove their influence. Pointedly, the masculinities stemming from my family entanglements were at work in the field even when my family was not physically present. Simply by being a father and a husband I was afforded a particular vantage over what was going on, and once my family status had been established in the field, my masculine positions afforded a resource for intersubjective discussion. Not only did my husband-father-hood afford me entry points for discussions, providing common ground for subjects to relate to me as something other than a researcher, it also afforded me a common ground from which *I* could relate to the subjects in my field. People would often ask me how things were going with my family and with my work, and I could ask them about *their* families and *their* work too: moreover, the common bridge of similar family positions afforded mutual insight and understanding about *what* was being said. In fact, I often used my masculine positions as common ground not only to build rapport but also as a verbal 'affordance' (Gibson, 1977) to help me plumb the entanglements of enterprising and family in my discussions; moreover, my entanglements of husband-fatherhood afforded a *mutual* 'resource' enabling meaning-making in intersubjective discussions (Holstein & Gubrium, 1995; Warren, 2012). In this way, various *disembodied presences* of my family materially influenced my research in the field, not only as traces shaping what I could discuss but as integral components of the masculinities-infused gendered self I carried into the field – part and parcel of my flawed and biased human research apparatus (Haraway, 1988). As Silverman (2014: 39) cautions, 'the facts we find "in the field" never speak for themselves but are impregnated by our assumptions'. In sum, our assorted gender positions (masculinities and femininities) are inherent to our assumptions, not only shaping what transpires in the field but also what the researcher *can* apprehend.

As Frohlick (2002) found, the gendered positions that go with accompanied research involving families can be a resource facilitating fieldwork, enabling interactions and revealing aspects of the field that might otherwise remain obscure. I cannot but agree. Contours of lifestyle enterprising have been revealed to me as a result of the masculine positionalities variously activated by my entanglements of family and field. As I see it, the influence of these entanglements extends beyond the matter of embodied co-presence of family in the field, and goes beyond even the signification of its traces in the externally attributed gender roles attaching to my appearance as a middle-aged married man wearing a wedding ring, to encompass the assorted masculinities that are indwelling *in* me, part and parcel of the inherently biased research apparatus of my gendered human self (Haraway, 1988). My contingent masculinities are variously active in my whole life, including in those moments when I enact my professional role as researcher both in and out of the field. Thus, the masculinities of father-hood and husband-hood are part of my research because, whether my family are with me in the field or not, they are not absent but remain 'present' *in me*, shaping my whole relation to the world. For researchers who have families, the masculinities arising in the intersection of family and field are always *essentially present*.

The Field in Family

Looking from the other side, my father-husband-hood was impacted by my researcher-hood in the place I called 'the field'. For my children, my field trips were simply family outings, visiting interesting places with mummy and daddy, and for the most part they were happy enough just to be doing something with us. However, like all of us, their enthusiasm for outings ebbed and flowed and here their individual temperaments came into play. My daughters have always been eager to go out and explore, so they were usually happy to be visiting the places that were my field sites. Their enthusiasm afforded me the impression that in bringing them to the field I was also discharging my fatherly duties to give them 'family time'. Being more of a 'homebody', my son was typically less enthusiastic about these field trips. Even though he did seem to enjoy these outings when we got to wherever it was we were going, he often seemed reluctant to go. Occasionally, his reluctance was such that I wondered if I was inflicting these outings on him, my researcher-hood coming at the expense of my fatherhood duty to prioritise his needs as my son.

Positional conflicts took hold in my husband-hood as well. There is simply a lot for parents to do when it comes to looking after children, and the task of organising and managing children to partake in any activity brings its share of challenges. Perhaps this was why my suggestions for such family field trips sometimes met with apprehension on the part of my wife; when her apprehensions proved well-founded, her harried visage served to remind me that the work of family care was gendered (Small, 2005; Wylie, 1987). Her distress revealed the blurring of my personal and professional domains and highlighted that my mixed visitation purposes had amplified her care-work burden. In particular, although she did not see the places we visited as 'the field', she was aware that such places were *my field* and that I was visiting these sites to also perform research and so, despite my reassurances, she made even greater efforts to manage the children to minimise potential disruptions. In this way, my dual role incidentally inflicted a dual role on her too, adding the burden of caring for my work as a researcher to the care burden she bore as wife-mother. When events transpired in such a way that she was visibly worn out by her dual burdens, the unsuccessful mixture of my professional and personal domains also challenged my masculine ideals, signalling failures as husband and father. These experiences suggest to me that the entanglement of personal and professional domains in familial field accompaniment not only affects the *researcher* in the field but also reverberates in the personal domain of family. If nothing else, this serves to highlight that the gendered burdens at the intersection of family and field come in many different guises, and that could be something researchers with families might need to be mindful of.

Concluding Remarks

Various masculinities arise at the intersection of family and field, taking hold both in the personal domain of family and in the professional domain of the field. The researcher-hood of my profession has at times spilled into my family, challenging my masculine ideals of father-husband-hood; at other times my masculine ideals of husband-fatherhood have spilled into the professional domain of my researcher-hood, shaping my relation to the field. In the intersection of family and field, tidal forces of relocation-dislocation blur the boundaries of family and work, their effects reverberating in personal and professional domains. The forces of relocation-dislocation play out differently between the researcher and their accompanying family and among individual family members; the uneven distribution of benefits and burdens reveals gender positionalities and normative structures arising at the intersection of family and field.

In the professional domain, family influences the field. Feminine vantages have highlighted that family positions affect the enterprises of research before and after the field, influencing such things as choice of research topics and choice of fields as well as the work that happens in the field (e.g. Frohlick, 2002; Khoo-Lattimore, 2018; Korpela *et al.*, 2016; Porter, 2018). My experience suggests the same can be said from the masculine vantages of father-husband-hood. Furthermore, these influences do not start and stop with co-presence in the field, but extend to 'absent presences' of family members as well. For researchers with families, family influence is never 'absent' because the gender positions that go with family are always *present* in the inherently flawed and biased human apparatus that the researcher carries into the field (Haraway, 1988).

For my part, assorted contingent masculinities stemming from my family positions of father-hood and husband-hood have affected my whole research enterprise, not only *in* the place I call 'the field' but also in the spaces that lie 'before', 'after' and 'beyond'. From this vantage, the assorted 'presences' of family variously arising in accompanied and unaccompanied settings, as observable or unobservable manifestations of masculinities in husband-hood and father-hood, cannot but leave their marks upon the whole enterprise of my research. The erasure of families from fieldwork accounts is entirely untenable (Frohlick, 2002; Korpela *et al.*, 2016).

Moreover, the entanglements of family and field illuminate the blurriness of professional and private domains, unpicking the illusory fabric separating 'work' from the rest of our living. For people who are sometimes researchers, this also serves to debunk the myth of research as some reified activity separable from the existential epistemology of gendered social life. In this, I would suggest that all researchers are personally entangled in gender-riddled processes of living, including those bits that happen when performing the academic occupational role of 'researcher', whether in the field or not. That assorted masculinities and femininities affect the research enterprise should come as no surprise. As Herod (1993: 306–307) says: 'Given that we live in a society in which gender relations significantly shape the lives of women and men, we should expect that such gender relations also shape the research process.'

So while my reflections have focused on the masculine entanglements of family and field, the wider message I wish to convey is that of the unavoidable 'presence' of gender in everything we do. As an existential epistemological condition of human social life, those people who are sometimes researchers have little choice but to tackle gender everywhere, in *and* out of the field. In this enterprise, the reflective researcher is doubly troubled, both by the problem of finding gender *in* the field and by the problem of finding where the entanglements of gender and field start and stop. Additionally, the trouble may be tripled for men, like me, who face the extra problem of gender blindness in the insidious inclination to perceive a genderless world (Kimmel, 1993). The latter trouble possibly helps explain why gender has remained marginal in tourism research (Figueroa-Domecq *et al.*, 2015) and the lack of reflexive accounts of gender in the field, particularly from masculine vantages (Porter & Schänzel, 2018). Pointedly, it seems as if the illusion of scientific objectivism stands obdurately in the way of gender, like the fabled statue of Ozymandias in the poems of Shelley and Smith, an assumption of genderless-ness (Kimmel, 1993) sustaining the myth of androgyny in tourism fieldwork (Porter, 2018).

However, there is no Archimedean point (Haraway, 1988), and the complexity of gender offers no remit to avoid its difficulties. The problem of gender will not just go away if we close our eyes to it, so there is no option but to face it. In tackling this trouble, the reflexive researcher is called to a wider accounting of gender in their relation *to* the field, including the bits that lie 'beyond', 'before' and 'after' the spaces of the field. What is called for in this troubled enterprise is an ethos of unabashed transparency, revealing our gendered selves both to ourselves *and* to those who are subjected to the knowledge that we publicly claim. If the core of research enterprise is to advance our knowledge of the social world, we must all work to find gender in the field *and* reveal its many presences to others in print (Porter & Schänzel, 2018).

Note

(1) 'The world is your oyster' is a quote from Shakespeare's play, *The Merry Wives of Windsor*.

References

Ateljevic, I. and Doorne, S. (2000) 'Staying within the fence': Lifestyle entrepreneurship in tourism. *Journal of Sustainable Tourism* 8 (5), 378–392. doi:10.1080/09669580008667374

Berglund, K., Ahl, H., Pettersson, K. and Tillmar, M. (2018) Movi(e)ing practices of gender, rurality and entrepreneurship. Paper presented at the Third Annual Entrepreneurship-as-Practice Conference and PhD Consortium, 16–20 April, Linnaeus University Växjö, Sweden.

Burns, P. (2001) *Entrepreneurship and Small Business*. London: Palgrave Macmillan.

Butler, J. (2004) *Undoing Gender*. London and New York: Routledge.

Canosa, A. (2018) 'Mummy, when are we getting to the fields?': Doing fieldwork with three children. In B.A. Porter and H.A. Schänzel (eds) *Femininities in the Field: Tourism and Transdisciplinary Research* (pp. 84–95). Bristol: Channel View Publications.

Carlsen, J., Morrison, A. and Weber, P. (2008) Lifestyle oriented small tourism firms. *Tourism Recreation Research* 33 (3), 255–263. doi:10.1080/02508281.2008.11081549

Cederholm, E.A. (2015) Lifestyle enterprising: The 'ambiguity work' of Swedish horsefarmers. *Community, Work & Family* 18 (3), 317–333.

Cederholm, E.A. and Åkerström, M. (2016) With a little help from my friends: Relational work in leisure-related enterprising. *The Sociological Review* 64 (4), 748–765. doi:10.1111/1467-954X.12375

Cederholm, E.A. and Hultman, J. (2010) The value of intimacy – negotiating commercial relationships in lifestyle entrepreneurship. *Scandinavian Journal of Hospitality and Tourism* 10 (1), 16–32.

Connell, R.W. (2005 [1995]) *Masculinities* (2nd edn). Cambridge: Polity Press.
Cornet, C. (2013) The fun and games of taking children to the field in Guizhou, China. In S. Turner (ed.) *Red Stamps and Gold Stars: Fieldwork Dilemmas in Upland Socialist Asia* (pp. 80–99). Copenhagen: NIAS Press.
DeWalt, K.M. and DeWalt, B.R. (2010) *Participant Observation: A Guide for Fieldworkers* (2nd edn). Lanham, MD: Rowman Altamira.
de Wit Sandström, I. (2018) *Kärleksaffären: Kvinnor och köpenskap i kustens kommers*. Göteborg and Stockholm: Makadam.
Di Domenico, M. and Lynch, P.A. (2007) Host/guest encounters in the commercial home. *Leisure Studies* 26 (3), 321–338. doi:10.1080/02614360600898110
Farrelly, T., Stewart-Withers, R. and Dombroski, K. (2014) 'Being there': Mothering and absence/presence in the field. *Sites: A Journal of Social Anthropology Cultural Studies* 11 (2), 25–56.
Figueroa-Domecq, C., Pritchard, A., Segovia-Pérez, M., Morgan, N. and Villacé-Molinero, T. (2015) Tourism gender research: A critical accounting. *Annals of Tourism Research* 52, 87–103. doi:doi:10.1016/j.annals.2015.02.001
Frohlick, S.E. (2002) 'You brought your baby to base camp?': Families and field sites. *The Great Lakes Geographer* 9 (1), 49–58.
Gibson, J.J. (1977) The theory of affordances. In R. Shaw and J. Bransford (eds) *Perceiving, Acting, and Knowing: Toward an Ecological Psychology* (pp. 67–82). Hillsdale, NJ: Lawrence Erlbaum.
Goffman, E. (1956) *The Presentation of Self in Everyday Life*. Edinburgh: University of Edinburgh.
Haraway, D. (1988) Situated knowledges: The science question in feminism and the privilege of partial perspective. *Feminist Studies* 14 (3), 575–599. doi:10.2307/3178066
Herod, A. (1993) Gender issues in the use of interviewing as a research method. *The Professional Geographer* 45 (3), 305–318.
Holstein, J.A. and Gubrium, J.F. (1995) *The Active Interview*. Thousand Oaks, CA: Sage.
Hultman, J. and Cederholm, E.A. (2010) Bed, breakfast and friendship: Intimacy and distance in small-scale hospitality businesses. *Culture Unbound: Journal of Current Cultural Research* 2 (3), 365–380.
Khoo-Lattimore, C. (2018) The effect of motherhood on tourism fieldwork with young children: An autoethnographic approach. In B.A. Porter and H.A. Schänzel (eds) *Femininities in the Field: Tourism and Transdisciplinary Research* (pp. 126–139). Bristol: Channel View Publications.
Kimmel, M.S. (1993) Invisible masculinity. *Society* 30 (6), 28–35. doi:10.1007/bf02700272
Korpela, M., Hirvi, L. and Tawah, S. (2016) Not alone: Doing fieldwork in the company of family members. *Suomen Antropologi* 41 (3), 3–20.
Kusenbach, M. (2003) Street phenomenology: The go-along as ethnographic research tool. *Ethnography* 4 (3), 455–485.
Levey, H. (2009) 'Which one is yours?': Children and ethnography. *Journal of Qualitative Sociology* 32 (3), 311–331. doi:10.1007/s11133-009-9130-8
Marcketti, S.B., Niehm, L.S. and Fuloria, R. (2006) An exploratory study of lifestyle entrepreneurship and its relationship to life quality. *Family and Consumer Sciences Research Journal* 34 (3), 241–259.
Morrison, A.J. (2006) A contextualisation of entrepreneurship. *International Journal of Entrepreneurial Behavior & Research* 12 (4), 192–209.
Peters, M., Frehse, J. and Buhalis, D. (2009) The importance of lifestyle entrepreneurship: A conceptual study of the tourism industry. *Pasos* 7 (2), 393–405.
Porter, B.A. (2018) Early motherhood and research: From bump to baby in the field. In B.A. Porter and H.A. Schänzel (eds) *Femininities in the Field: Tourism and Transdisciplinary Research* (pp. 68–83). Bristol: Channel View Publications.

Porter, B.A. and Schänzel, H.A. (2018) Introduction. In B.A. Porter and H.A. Schänzel (eds) *Femininities in the Field: Tourism and Transdisciplinary Research* (pp. 1–9). Bristol: Channel View Publications.

Schänzel, H.A. and Smith, K.A. (2011) The absence of fatherhood: Achieving true gender scholarship in family tourism research. *Annals of Leisure Research* 14 (2–3), 143–154. doi:10.1080/11745398.2011.615712

Shaw, G. and Williams, A. (1987) Firm formation and operating characteristics in the Cornish tourist industry – the case of Looe. *Tourism Management* 8 (4), 344–348. doi:10.1016/0261-5177(87)90092-6

Silverman, D. (2014) *Interpreting Qualitative Data* (5th edn). London: Sage.

Small, J. (2005) Women's holidays: Disruption of the motherhood myth. *Tourism Review International* 9 (2), 139–154. doi:10.3727/154427205774791645

Sultana, F. (2007) Reflexivity, positionality and participatory ethics: Negotiating fieldwork dilemmas in international research. *ACME: An International E-Journal for Critical Geographies* 6 (3), 374–385.

Swanson, S.S. (2018) The married life (as a marine tourism researcher). In B.A. Porter and H.A. Schänzel (eds) *Femininities in the Field: Tourism and Transdisciplinary Research* (pp. 37–52). Bristol: Channel View Publications.

Tourism Skåne (2019a) *About Skåne*. See https://visitskane.com/about-skane.

Tourism Skåne (2019b) *Visit Skåne: Southeast*. See https://visitskane.com/cities-locations/southeast.

Tourism Skåne (2019c) *Visit Skåne: Southwest*. See https://visitskane.com/cities-locations/southwest.

Warren, C.A.B. (2012) Interviewing as social interaction. In J.F. Gubrium, J.A. Holstein, A.B. Marvasti and K.D. McKinney (eds) *The SAGE Handbook of Interview Research: The Complexity of the Craft* (2nd edn) (pp. 129–143). Thousand Oaks, CA: Sage.

West, C. and Zimmerman, D.H. (1987) Doing gender. *Gender & Society* 1 (2), 125–151.

Wylie, J. (1987) 'Daddy's little wedges': On being a child in France. In J. Cassell (ed.) *Children in the Field: Anthropological Experiences* (pp. 91–120). Philadelphia, PA: Temple University Press.

13 Fatherhood in the Field: Reflections on Kinship, Identity and Ethnographic Research

Michael A. Di Giovine

Introduction

A few days before leaving for the field to conduct two years of dissertation research in Italy, one of my professors at the University of Chicago, Paul Friedrich – a pioneer of humanistic anthropology and the study of 'peasant' cultures and local politics in Mexico and the Slavic world (see Friedrich, 1977) for whom I served as his research assistant at the time – took me aside and gave me some sage advice: 'When you're in the field, make sure to have children.' While at first this seemed a bit of an oddly personal statement, he knew I was leaving with my wife and that we had been contemplating starting a family; in fact, those considerations had been formative in my switch from researching UNESCO World Heritage sites in Southeast Asia (see Di Giovine, 2009) to pilgrimage and development in the town in Italy outside of which my mother was born (Di Giovine, 2012a). On the one hand, my wife preferred to live (and possibly start a family) in Italy rather than Cambodia, as she was born to American expats in Rome, and on the other hand, I had also become interested in the changes brought about by pilgrimage and heritagisation in the small town during my family visits, since the canonisation of 'our' saint, Padre Pio of Pietrelcina, who had become, as one journalist put it, 'the world's most popular saint' (Wilkinson, 2008). Friedrich continued: 'I did it. If anything, it's good for the research. It humanises you, makes you approachable. People will talk to you.' As if on cue, a week after we arrived in Italy we discovered my wife was pregnant with our first son, beginning an unexpected (although not entirely unplanned) journey across fatherhood and kinship networks, health and immigration systems, religious practices, race and identity. Becoming a father – as well as being a son, husband and kinsman – opened doors for me; my positionality gave me great insights into the deeper, more hidden areas of Italian social life that, while on the surface seeming tangential to tourism and pilgrimage research, shed light on the broader social and cultural context of those living and travelling through the land of Padre Pio.

Part autoethnography and part reflection, this chapter presents a series of ethnographic vignettes to illuminate the role fatherhood and families play in conducting research in the field, as well as the craft of ethnographic inquiry more generally. Ethnography is the set of qualitative methodologies such as participant observation, interviewing, kinship mapping and oral history elicitation that is part and parcel of anthropological research. Acting as both a participant and an outside observer, the anthropologist walks a fine line between subjectivity and objectivity to gain visceral and embodied insight into his or her subjects' emotions, motivations, dispositions and worldview. Based on long-term ethnographic fieldwork on pilgrimage and tourism in southern Italy, this chapter argues that experiencing fatherhood – as a dad, son and grandson – as well as being a kinsman and husband, can provide illuminating insights in this type of research: for the former, it both humanises and allows for living the sort of lifecycle experiences only natives experience in the site; for the latter, it allows for the researcher to call on premeditating kinship networks, alliances and rivalries that could provide unexpected entrée into the private and public lives of locals.

Ethnography and Anthropological Positionality

The hallmark of anthropological practice, ethnography is the bundle of qualitative research methods that aim to elicit data concerning the motivations, behaviours, rituals, worldviews and other 'social facts' (Durkheim, 1982 [1895]: 74) of members within a given social group; it adds depth and dimension to the quantitative research that has particularly dominated interdisciplinary tourism research and much of the sciences. Predicated primarily on participant observation – that is, living and participating in daily life activities with locals while also observing, documenting and, later, analysing them – and supplemented by ethnographic interviews that are free-flowing, goal-oriented conversations – ethnography is an emic or 'bottom-up' approach to data collection; it values locals' agency in determining the nature of their contact with the researcher, focuses on lived experiences and elicits questions and answers that emerge from the research participants rather than those predetermined by the researcher. As the primary purpose of ethnography is to 'grasp and then to render' (Geertz, 1973: 10) the ways in which individuals and social groups behave within a culture, ethnography helps to shed light on the latent 'imponderabilia of daily life' (Malinowski, 1984 [1922]: 18–20). That is, it reveals norms and values that are often implicit, normalised, taken-for-granted and unarticulated by the subjects themselves, but which shape and inform their everyday behaviours, attitudes and decisions. Being emic does not mean that anthropologists enter the field blindly: we prepare for the field through intense study of the field site's culture, history and environment; we must speak the language of the locals in order to understand the minutiae of their discourses in their own words; and we must familiarise ourselves with theories and pre-existing research on the specific topic of study. Although shaped by our experiences in the field and our daily interactions with locals, as well as serendipitous 'ethnographic moments' brought on by an unexpected encounter or experience which would suddenly clarify or illuminate an important facet of the research, we therefore nevertheless employ a particular gaze on 'ethnographic things' (Stoller, 1999). In that way, we prepare ourselves to concentrate on those things germane to our research – otherwise we might be overwhelmed by the cacophonous complexity of everyday life.

While ethnographers enter the field with a somewhat precise idea of whom they wish to study, and perhaps even a few interview subjects already lined up, it is through participating in daily life that certain subjects, experiences and events are discovered. 'Key informants' are typically found in the field; they often become trusted conversation companions, helping the ethnographer navigate an oft-unfamiliar world, and introducing the researcher to people, places and events that the informant – as a member of that society – considers to be most illuminating. This process is iterative; an ethnographer may be introduced to one person, and that person may then provide insights and contacts that the researcher will use to delve more deeply into the culture, and to gain more subjects. Such 'snowball sampling' may be looked down upon by top-down or etic researchers for its seeming lack of rigorous preplanning, its fluidity, its inherent subjectivity and the self-selective nature of the subjects. But as a qualitative emic science, anthropology values quality over quantity, illuminating narratives over controlled representationalism, deep engagement with a limited number of revelatory subjects over sample size. Indeed, although quantitative sciences may seek to project objectivity through anonymous surveying, modern anthropology recognises the inherent subjectivity of social science research and has become quite reflexive concerning the positionality of the researcher, even embracing it as I have. For example, rather than remaining coolly dispassionate on pilgrimages, going through the motions perhaps but attempting to remain 'neutral', I pray with devotees, share my stories and listen to theirs. In a word, I am a participant, and this has helped foster trust and sincerity among my research subjects, allowing them to open up to me and even to provide what they see as assistance to my own growth. 'Here', one devotee once told me, pulling out a plastic rosary from her purse unannounced, 'Keep it; I can see you need it'. Such experiences provide an opening for engaging in deeper conversation concerning spirituality, her devotion and the need for prayer.

Since ethnography is an interpretative, emic science which is predicated on personal experience and relationships, it would seem obvious that the researcher's social status – including his sex, kinship status and personal networks – would influence not only his analysis but also, more importantly, the ethnographic gaze employed (cf. Comaroff & Comaroff, 1992). But as this chapter shows, it also conditions the very interpersonal encounters of fieldwork, often providing unplanned and rather unexpected moments that definitively shape the research (see, for example, Strathern, 1999: 3–11). My fieldwork would have been very different if I had a different status or social networks – and indeed no two fieldwork experiences are alike. Yet it was only during the post-modern 'reflexive turn' beginning in the 1970s – which was tied to humanistic, feminist and, later, queer anthropology and developed in the 'writing culture' debate of the 1980s and 1990s (Clifford & Marcus, 1986) – that ethnographers acknowledged the fact, focusing on the 'anthropology of experience' (Turner & Bruner, 1986).

Gender has always played an important role in colouring the ethnographic experience, as well as its data and analysis. Prior to the reflexive turn, ethnographic data – most often collected and interpreted by white, upper-class males – were presented unproblematically as cultural fact, frequently minimising the roles of women, minority groups and outsiders such as tourists (Handler, 2010; Lutz, 1990; see also Handler, 2000). The male-centric words of Bronislaw Malinowski, the 'father of ethnography', are telling: ethnography's ultimate mission in presenting a holistic view of the native subject was 'to realize *his* vision of *his* world' (Malinowski, 1984 [1922]: 25; *emphasis in the original*). This was a convention of the time: to remain scientific, careful erasure of the observer was practised, even though the data were generated from subjective, personal encounters and therefore revealed something of the ethnographer's personal experience (see Collins & Gallinat, 2010: 2). Furthermore, masculinity was an unmarked category; it was taken for granted. Kinship, familial and sexual relations were told largely from the male perspective, although sometimes data on women or children would be collected by the anthropologists' wives or children who accompanied them to the field. Yet those relations, who often were not formally trained in ethnography, were largely left unacknowledged – as was the case with the early work of Victor Turner and others (Edith Turner would formally co-author their work on pilgrimage (Turner & Turner, 1978), before becoming recognised as an anthropologist in her own right after his death (see Engelke, 2004)). Early exceptions were Beatrice and John Whiting's (1975) *Children of Six Cultures* study on child-rearing and Yolanda and Robert Murphy's co-authored monograph, *Women of the Forest* (1974), which examined the patrilineal, matrilocal society of the Amazonian Mundurucú whose worldview was predicated on the mythological animosity between the sexes. Feminist anthropology pioneered by Sherry Ortner (1974), Marilyn Strathern (1988), Henrietta Moore (1988) and others paid attention to the role of gender in shaping the ethnographic experience and reinserting women in anthropological narrative. Queer theory did the same for sexuality (the collection 'Queer Futures' edited by Boellstorff and Howe (2015) reflects on how one's sexuality conditions the research irrespective of the study's theme). Barbara Myerhoff's (1978) study of elderly residents of the Jewish ghetto considers the role of religion and age as well. Finally, Ester Newton's classic, 'My best informant's dress' (1993), which recounts her love affair with her primary research subject, an elderly woman, fuses all of these; she shows how gender, sexuality, age and even class create the unique conditions, and insights, for her research.

It is in this vein that this chapter is written. Here, I reflect on the ways in which my long-term ethnographic fieldwork on Padre Pio pilgrimage, as well as my current research on sustainable food and cultural heritage in central Italy, has been uniquely shaped by my own kinship (both affective and achieved), sex and gender and, of course, fatherhood. The vignettes that follow each address a particular status and how that status illuminated a key 'social fact' in my research: from affective kinship (the kin system into which one is born), to achieved kinship (one that is earned), to being a father, son-in-law, citizen and *paesano* (fellow townsman).

Affective Kinship, Social Networks and Identity

Before even becoming a father in the field, I was first and always a kinsman – a son, grandson, cousin, nephew and *paesano* (literally 'townsman' or 'countryman', but also meaning a fellow Italian-American back at home in the United States). Kinship, I found, was particularly important for devotional practices in Pietrelcina. Pietrelcina was the birthplace of Padre Pio, and although his famous ministry – marked by supernatural occurrences such as ecstatic visions, bilocation (being in two places at once), knowing one's transgressions before they confessed them and, most famously, 50 years of the stigmata (the bleeding wounds of Christ's crucifixion on his hands, feet and side) – took place in the town of San Giovanni Rotondo some 160 km away, Pietrelcinesi entertain a close relationship with him, their unique devotion predicated on their familial ties to Pio and his family rather than on the kind of transactional saint-devotee relationship traditionally expressed towards a town's patron saint (Di Giovine, 2012c). Even if they did not trace their blood relations to him, most Pietrelcinesi nevertheless have some story of a personal relationship or encounter he had with their family, and I am no exception. It is said that my grandfather, who was born in the town some 20 years after Pio, shared the same godfather with the saint. While it is not possible to verify this, as my family's knowledge-bearers from that era had passed away (and he died shortly after emigrating to the United States when my mother, his youngest daughter, was only 12), I was eager to rationalise why I, this 'American', was living there by tracing such lineage.

My identity as a kinsman was not always evident to locals (especially since I trace it matrilineally and therefore it is not reflected in my last name) and, when asserted, I was often made to perform this connection to them. Identity itself is a social construct; it is a performance, a way to present to oneself and to others one's affinity with a group through the oft-selective deployment of stereotypes, proper norms and behaviours befitting of the status, and close-knit relationships (see Barth, 1956). The most expedient way was to provide people with one's last name. Small-town Italians are especially sensitive to such clannish identity and, as families cluster primarily within particular villages, are quite adept at identifying with startling precision one's provenance; at the least, many can identify the region one is from by how the name is spelled or pronounced. Particularly in patrilineal societies – of which identification with a last name is a remnant – tracing one's descent group is important for forging alliances such as marriage, understanding whom to trust or call upon in times of difficulty, and informing proper social behaviours. Traditionally this was important in southern Italy, where society, argued Edward Banfield (1958) in his well-known ethnography of an impoverished, post-war southern Italian town, was dominated by an ethos of 'amoral familism' – an action was just if it protected or enhanced the wellbeing of the family above all else, and therefore mistrust towards those outside the family was rampant. While Banfield's analysis of what he called a 'backwards society' smacked of the kind of exoticising ethnocentrism prevalent at the time, and could have been a reaction to his own positionality within the town (much of the data were collected by his wife, who was from the village and spoke the dialect, yet she remained largely uncredited in the work; Laurino, 2000: 39), Italians at the very least do celebrate what they affectionately call *campanilismo* – identifying above all with their town or neighbourhood's bell-tower (*campanile*). This practice informs their notions of authenticity, proper behaviour and even prescribed temperament. Indeed, this strong tie to the particular place, which is conceived of as bestowing on the person his or her essence, is often conceived and talked about as a sort of human *terroir* (to use alimentary terminology denoting the characteristic flavour of the land from which a food or wine was produced). People will use the appellation DOC which denotes a wine's terroir discursively for expressing one's authenticity: just as a wine from the Sangiovese grape can only be designated as Chianti DOC if it is grown and processed in the Chianti region – as it produces the unique 'taste of place' (Trubek, 2008) which is part biological (i.e. the unique mineral content of the particular field, grown on a particular hillside under particular atmospheric and solar conditions) and part social (i.e. the grape is made into wine through the traditional process of the area) – Italians will say a person is '*Pietrelcinesi* DOC' if he has the correct pedigree, expresses the correct stereotypical temperament, cultivates the correct taste preferences for the food, and reasons in the same stereotypical ways as others in the town; the person, in essence, is seen as a genetic and geographical product of the land (Di Giovine, 2015a: 83–84).

The problem is, I am no Chianti, so to speak.

My mother's maiden name was not from Pietrelcina but from the next town over, Pago Vaiano, although we have no ties there. My grandfather's ancestors must have been immigrants to Pietrelcina. Consequently, when I would assert that my family was from Pietrelcina and I was asked my grandfather's last name, I would inevitably get a reaction akin to 'stupid American; he doesn't even know what town he's from!'. I would push back, explaining precisely which house was his (although this was sometimes contested), who else were his kinsman and, finally, I learned, with his *soprannome* ('nickname'). While the passing down of last names is an important residual aspect of an early patrilineal society, Italians (like most Euro-Americans) nevertheless trace kinship bilaterally (that is, they see themselves related to both mother's and father's sides equally), and so bonds among extended family can become weak and diluted. I found that, since villages were dominated by only a few big family names, such that the town is made up of family members far removed from any meaningful lived relationships (i.e. they are fourth and fifth cousins and therefore do not see themselves as closely related), they would go instead by a *soprannome*. My grandfather's nickname was *mozzon'*, or 'shorty' in the Pietrelcinese dialect. Once I would say this, locals would accept me: 'Oh, now I know who he is!'. This would then initiate frank conversations on life in early 20th century Pietrelcina, family networks and locals' devotion to Pio.

Kinship networks are particularly important for subsistence in southern Italy, where locals – as Banfield also noticed – exhibit strong mistrust or ambivalence towards formal institutions (Banfield, 1958; see also *Corriere della Sera*, 2018). This had been very frustrating for early Italian statesmen who, attempting to foster a sense of national identity as well as economic development, could not make significant headway in the south. This *'questione meridionale'* (the 'Southern Problem'), as Antonio Gramsci (1926) euphemistically called it, stemmed from a long history of aristocratic domination beginning with the Roman latifundia system and the unique organisation of the medieval *chiese ricettizie* (churches of received local priests, who were more independent from Bishops than in the north) which frustrated the Tridentine Church's attempted changes during the Counter Reformation (Carroll, 1992: 97). The *questione meridionale* remains a salient discussion as much today as it did for Gramsci, where infrastructural development – including that earmarked for tourism – seems to not be effective no matter the amount of funding provided (see Schneider, 1998). And indeed, prior to the boom in religious tourism at Pietrelcina, the region was one of the poorest in Italy (Davis, 1998).

The importance of traditional kinship networks became clear during my first real ethnographic moment. As soon as my wife and I arrived in Italy, we stayed with a priest friend in the north while my uncle, a self-sufficient farmer in the southern countryside outside Pietrelcina, helped procure living arrangements for us. Despite my insistence, he would not take me to an *immobilare* ('real estate agency'), saying they could not be trusted. Rather, the baker from Pietrelcina who would deliver bread twice a week had a relative who was the postman in town and who would rent his basement to us. I visited the space with my uncle and the postman, who told me he would fix it up for us. He promised to install the *cucina* ('kitchen') for me, since in Italy when people move they typically take all appliances with them (the stove, refrigerator, cabinets – and literally the kitchen sink). I certainly did not want to buy one since my stay was only temporary. We agreed on a price and I left for a conference in the UK. Upon my return back to my wife in the north, I received a call from my uncle saying the apartment was ready: 'Everything's ready; you just have to buy the *cucina* and you'll be set', he said. I balked. The postman had clearly said he would provide it; my uncle heard him too. Yet my uncle would not admit (nor deny) this 'bait-and-switch', instead trying to convince me that I needed to buy the kitchen because it would be socially disastrous for us if I didn't. The postman was of a higher status than he was, and it was important not to burn bridges in his limited social network.

I called my great-uncle, who had always been like a grandfather to me and who was a retired general in the carabinieri (military police). His wife was from one of the five prominent families in Pietrelcina, and she put me in touch with a cousin on her side who was married to a *finanziere* (finance police, akin to the American FBI). They had a tiny *mansarda*, or attic apartment, that they rented to me (with a kitchen included!). We took it. My uncle on the farm was mortified as I had forced him to lose face with members of his network, and despite our closeness it took many months to resolve.

There is a saying in Italian that the only person more powerful than the priest in a town is the *finanziere*, and I found this to be true. Instead of being harassed by the postman or shunned by locals as I had feared, we were now under the protection of one of the most powerful families in Pietrelcina. But most importantly, I was now part of the *finanziere*'s social network: on the street, people would ask if I was the 'American' who was the cousin of his wife's cousin; they would talk to me, welcoming me into their cafes, restaurants and tourist shops. The *finanziere* would take me around the town, organising meetings with his powerful friends such as the mayor and the *assessore* (the town council member) and the 'professors' (educated school teachers). My wife and I participated in closed devotional events reserved only for locals, such as preparing the sacred statue of the Madonna della Libera, the town's patron saint, for her annual festival and procession. I was even interviewed on local television as the 'American professor'. In this way, not only did the difficult negotiation of kinship identity, status and social networks shed light on deeply engrained cultural practices, but it also facilitated my entrée into the very closed world of local kin networks that was only accessible to 'authentic' Pietrelcinesi.

Being a Husband: Race, Gender and Citizenship

The first step in preparing ourselves for having a child in Italy was to make sure that my wife was covered by the national healthcare system that was granted to all citizens. Although she was born in Rome and lived there for the first few years of her life, my wife's parents were Americans working in the US Embassy and therefore she was not legally considered Italian. Unlike the United States, which grants citizenship based on the principle of *jus soli* ('law of the soil' or birthright citizenship), that is, you are a citizen of the country in whose geographic territory you were physically born, Italy and much of Europe grant citizenship based on *jus sanguinis* ('blood right'), that is, whatever national 'blood' you have (i.e. the citizenship of your parents) makes you a citizen regardless of the territory in which you were born. This is particularly important for countries like Italy, from which nearly 6 million people emigrated between 1890 and 1920 and another half million after WWII (see Cavaioli, 2008; Femminella, 1961), but which is experiencing an overwhelming crush of new immigration from predominantly Muslim countries in eastern Europe (particularly Albania), northern and sub-Saharan Africa and the Middle East. These are largely unwanted migrants, not only as they are often impoverished and look, speak, worship, act and eat in palpably different ways, but because they also use Italy as a gateway to more economically powerful countries in northern Europe, who blame Italy for their inability to stop their flow into the European Union (see, for example, Cole, 1997).

Although my mother had to renounce her Italian citizenship when she naturalised in the 1950s, and my father was born in New Jersey to naturalised parents, I have dual citizenship through my father's side of the family – ironically the most Americanised side. This status not only grants me the *tessera sanitaria* (health insurance card) covering me under the socialised welfare system throughout Europe, but it gives me the right to vote in Italian and European Union elections; there are so many Italians abroad that, since 2001, 12 seats are allotted in the lower house of the Italian Parliament to represent our interests, and six in the senate (Law 459, 27 December 2001; see Ministero dell'Interno, 2020). So while culturally I am less Italian than the millions of North Africans who were born or raised in Italy – something that was made quite clear to me as a youth (Di Giovine, 2010: 181–182) – legally I have more rights than these 'foreigners'. I point out to my students, for example, that while I might look and legally be considered more Italian than someone like the great Italian soccer player Mario Balotelli, who was born in Sicily to Ghanaian parents, he is more Italian than me culturally. Yet he only obtained the same legal status as I have when he started winning. Short of being a national sports sensation, then, the most direct way for my wife to obtain these same rights is to marry a citizen.

Obtaining citizenship in Italy and elsewhere is a bureaucratic endeavour. In Italy at the time, the law stated that people must be married to an Italian citizen for two years if they live in-country, and three if they are married and live abroad. (In 2018, a language fluency test was added.) We had been married for two years and eight months when we moved to Italy; on our third anniversary we presented ourselves to the *comune* ('town hall') and requested the paperwork to begin the process. My wife was already five months pregnant. The official looked at our documents, including her birth and marriage certificate, and had a problem. Firstly, the name on her passport did not match that of the birth certificate; as is traditional in America (although this is changing), she had taken my last name. Secondly, the law clearly states that those living in Italy must wait two years to request citizenship. Technically, we were a married couple who had lived in Italy for only four months. No amount of rationalisation could make them understand the spirit of the law – that we had actually been married for three years, the maximum amount of time regardless of where one was living. The law was the law. We even appealed to the Prefect (the governor of the province), but it was futile. We would have to wait. What's a few more years? While this aspect sheds light on the role of bureaucracy in Italy for which the country is infamous – as well as perceptions of time in the country that pioneered the 'slow' movements (Slow Food, Slow Tourism) – it also underscored the reasons why many southern Italians are mistrustful of the government, preferring direct and unmediated appeals to more powerful members of their networks in a vast informal system of favours. Of course this is what the government would do, but *s'arrangia* – you get around it; you make do.

For us, it turned out that my status as an Italian husband, as well as a white American (the locally preferred type of immigrant) worked out well. Because my wife was married to me, and especially because she would be having my Italian baby, she was able to apply for residency and the *tessera* so she could give birth under the healthcare system. This process could take years, particularly for new migrants who did not speak the language and had no support network. But my friends and family members were police officers, and we could schedule meetings with the police's immigration office to expedite the service. If we could physically get through the door, that is. In the town (and, I have heard, other towns as well), one of the powers the police immigration office has over these undesired migrants is to lock them outside the building, every once in a while opening the door and magnanimously picking a few from the begging crowd to come in and wait for several more hours. We were usually let in because we had an appointment thanks to our friends, but once we were locked outside with the masses of Algerians, Nigerians and Albanians, knocking and begging to be let in. It started raining and my wife, whose baby bump was clearly evident, felt faint. The masses of migrants mobilised, banging louder and shouting that a pregnant woman was sick and needed to be let in. A police officer opened the door, looked at her and said, 'Why didn't you tell me who you are? You don't need to wait like them' – gesturing to the rest. Once inside, the officer clarified that in future we did not have to wait as I, her husband, was a (white) Italian by blood. Nor did we have to fill out forms and mail them to various offices; that was for the others.

The only real problem was my wife's last name. Because last names are key to understanding kinship alliances, ancestral origins and one's true *terroir*, Italian women keep their maiden names. No matter how intuitive it is to an American that someone born with the name Laura Anne Morrill who married a Di Giovine and now had a passport that said Laura Morrill Di Giovine was the same person, this seemed not to be evident to the Italians. We were even accused of incest! It was difficult to explain the American norms of marriage and naming, even to immigration officials whom we assumed would be more experienced with different cultures' naming norms. But race also came into play. On several occasions we would explain to an official that in America women take the name of their husbands; they would furrow their brows, sit back, and say, 'Oh, like the Arabs!'. 'Why would you want to do that, to lose your own identity like they do in the misogynistic, male-dominated Muslim world?', they would further ask, and often with such directness. We had no choice but to go to the US Embassy in Rome to sign an affidavit stating that my wife is the same person who is named on her birth certificate. When we met with the young consular officer, she smiled and said: 'Let me guess. You tried to do something official and they looked at your documents and said, "Who the hell are you?".'

I had lived in Italy for extended periods of time before, but my position as a husband and an expectant father had pushed us into situations that, I would argue, most ethnographers need not encounter. While fieldwork abroad is always subject to and shaped by politics – Malinowski conducted his Trobriand Island research because he was exiled there during World War I, and many others conducted theirs as colonial agents – the political situation of the researcher has also largely been ignored. Indeed, the history of anthropological fieldwork has been one of inequality – wealthy, educated and mobile European-American males living temporarily with largely impoverished, undereducated 'natives' with little ability to leave their geographic homes. But out of necessity because of having a child in Italy (a decision that we could afford to make as privileged Americans), we needed to actively engage with legal and political issues concerning citizenship and racial politics. This forced me to deploy my rights of Italian citizenship and, while ultimately successful thanks to my privileged position, it afforded us the possibility to understand viscerally the processes that many migrants are subject to. In the end, it shed light on the complexities of race and citizenship, bureaucracies and ideas of temporality, and legal and cultural belonging.

Firstfruits: Becoming a Father

Apart from adoption, fatherhood is typically an ascribed status. Unlike the achieved kinship statuses of adoption or marriage – in which some formal, socially sanctioned ritual occurs that acknowledges a voluntary status change – an ascribed status is simply assigned involuntarily by birth or through some event (such as childbirth). But this does not mean that it is intuitive. Since the role of fatherhood is culturally contingent, its expectations and values differing from society to society, one nevertheless must learn how to be a father. Like other forms of kinship it is a process, predicated on observing other fathers (starting with one's own), consuming media representations of fathers, participating in normative activities such as prenatal classes (which in Italy is the gendered term, '*pre-mamme*' classes), reflecting on these experiences and, of course, muddling along. Yet the act of learning to be a father abroad – from making life-decisions and compromises for the benefit of the child to mastering the norms and values of fatherhood – is a tension-filled act of acculturation (learning and attempting to master another culture), pitting the imaginaries, memories and understandings of fatherhood from one's birth culture against those learned in the field. Although stressful, it is also productive, as it highlights different cultural conceptions and worldviews that are often latent and unrecognised.

Fatherhood also forces you to make decisions that might conflict with your fieldwork plans. For example, about halfway through the pregnancy, my wife and I decided to transfer from Pietrelcina in the south to Pesaro to the north, where our priest friend lived. The decision was not taken lightly, but rather was predicated on two main factors. The first was tied to familial obligations and the burden we felt we had on my extended kin, who seemed to constantly need to perform rural norms of care and hospitality. However, cultural tensions also emerged, particularly around food and well-being. Subscribing to American norms of food safety, which state that pregnant women should avoid cured meats, unpasteurised cheeses, coffee and wine, my wife found it difficult and even insulting to deny the elaborate spreads of homemade *capicola* and *sopressata* salami, raw cheeses and wine from their vines. Explaining paediatricians' advice only elicited statements such as 'we ate it when we were pregnant and we were fine!'. Rationalising that cultural norms are different did not change this; nor did discussions about differing gut biomes.

Unfortunately, moving away from Pietrelcina was not met with approval, and *campanilismo* was in full display. This revealed deeper ideas of identity and personal terroir. Leaving Pietrelcina to give birth elsewhere meant having a child who was not Pietrelcinese. That we were moving to Pesaro, in the central Italian region of Marche, was particularly painful because Marchegiani are not looked on kindly by southerners; they were the papal tax collectors prior to unification. *Meglio un morto in casa che un marchigiano alla porta*, they say; better death in your house than someone from the Marche at your door!

I also was ambivalent about moving away not only because we seemed to be abandoning my kin, but also because I was worried about the integrity of my fieldwork. After all, the reason we came to Italy was to study Padre Pio pilgrimage and devotion in Pietrelcina. Moving nearly 500 km away – over a five-hour drive – would significantly impact my research, and I was prepared to all but put the study on hold until our son was born. Yet the decision proved serendipitous, for it allowed me to unexpectedly encounter different demographics of Italians who had different ideas of Padre Pio. Devotion to Pio was strong there, too; parishes made pilgrimages down to the south (in which I participated), devotees talked of their belief in the saint through different idioms, and even the sceptics, atheists and Protestants would discuss Pio, his importance and the difference between northern and southern Italian piety. Since this was also an extraordinary period in the cult of Padre Pio – the shrine had exhumed his body and put it on display for 18 months, drawing upwards of 6 million pilgrims – Pio was also on the television a lot. Religious officials, pop cultural icons, politicians and academics would be on talk shows such as *Porta a Porta* discussing the history and development of the cult, the veracity of his stigmata, the miracles attributed to him and the economic impact of the extraordinary veneration.

The most important reason for our move was the perceived difference in care. While health coverage was the same throughout Italy, the quality of the care – including the resources – differed between our small village (without a hospital) in the rural south and the middle-class town of Pesaro in central Italy – which Italian economists call the *terz'Italia* ('third Italy') – not quite the affluent and industrialised north, but not the impoverished south either. Rather, central Italy is marked by small and medium-sized businesses, a growing middle class and a mix of industrial and agricultural industry. It is also quite secular, and Pesaro is known as one of the *città rosse* ('red cities') for its proclivity towards communism over conservative Catholicism that marks much of southern Italy.

Despite its communist and areligious leanings, we found that Catholic values permeated many of the discourses justifying variations in care between the United States and Italy. For example, while in the United States an epidural is often covered by insurance – our Catholic hospital in Chicago called it a 'right' when we had our second son two years later – it was not covered by the socialised medical system in Italy and was viewed with a measure of disapproval. Even the most secular (female) friends would state that the pain of childbirth was a 'good' pain, given by God or nature. In true Italian fashion, however, there was a workaround: the expectant family would have a private interview with one of the few anaesthesiologists who could do the procedure, and then pay several hundred euros to the bursar in a semi-under-the-table operation. The problem, of course, was that there was no guarantee that the anaesthesiologist would be available when you went into labour, and another anaesthesiologist might not want to take you on as a client. This happened to us, but fortunately another Italian also had arranged for an epidural and so the substitute came in to administer the procedure.

Once my son was born, we found that the care did not stop. As in other countries in Europe, we were provided with a new parent gift box that included diapers, formula and rice cereal. A lactation consultant from the hospital visited about a week later to check on the status of the child. In line with Italians' embrace of herbal teas, aperitifs and digestives that accompany meals to aid in digestion, she said if he wasn't hungry or had colic to give him a chamomile *tisane* ('herbal tea'). When it was time to introduce solid food several months later, we found that differing cultural norms prescribed different types of first foods, even though these recommendations are often taken as universal norms. In Italy, our paediatrician advocated a little boiled meat, olive oil and Parmigiano cheese, while an Italian colleague who teaches in Estonia was told by paediatricians there to begin by introducing boiled potatoes. As our son developed his appetite, we could buy the same international brands of baby food as in the United States, but with a broader range of indigenous ingredients such as *cavallo* ('horse'), *cinghiale* ('wild boar') and my favourite protein, *coniglio* ('rabbit').

But most importantly, my son afforded us the kind of entrée into Italian society that Paul Friedrich had advocated. Walking him through the cobblestone streets, piazzas and seaside boardwalks, strangers would stop and touch our child (he was born with so much hair they would say it was like a baby wig!). When they heard we were American, their questions about why we had a child here would initiate conversations that I would steer towards my research question. He humanised us, made us interesting to talk to, and helped build trust among locals and us.

The Evil Eye

One of the first experiences that was directly relatable to my dissertation work on popular devotion occurred when we took our young son back to see my family in southern Italy. I had been struggling to better understand the differences between northern and southern Italian piety. While pilgrims from all over Italy flood Pio's shrine of San Giovanni Rotondo, their actions are different; northern Italians generally profess to being more reserved, preferring to pray to the saint, attend Mass and engage in the prayer groups that were set up by the saint in the 1950s and that provide an extensive worldwide network of devotion. In talking with locals in the northern city of Pesaro, they would often profess devotion to the saint but scepticism towards the 'exaggerated' and even superstitious way in which he is treated by southerners. I found that southerners exercise much more sensory and embodied devotional practices; they will touch and kiss statues of Pio, rubbing their rosaries and photos of loved ones onto anything that seems like a relic of the saint in the hopes of obtaining what Frazer (1952 [1922]: 43–52) called its 'contagious magic' (see Di Giovine, 2012b); northerners would also point out that southerners seemed to pray directly *to* Pio, rather than *through* Pio to God as is theologically proper. From my study of the literature, I knew that this was a historically situated activity; while Tridentine reforms in the 1500s Counter Reformation attempted to do away with these types of practices, which were deemed superstitious and which Martin Luther and Protestant sects criticised, the Renaissance Church made little headway in the south. As Carroll (1992) shows, while in the north the Church had direct power through Bishops to enact such re-education, southern Italy's alternative organisation into *chiese ricettizie* (churches of received priests who came from the local community and privileged local kinship alliances over the bureaucratic authority of the Pope and his Bishops) made reform particularly difficult. In explaining the *questione meridionale*, for example, Gramsci (1996: 495–496) mocked these southerners for their longstanding superstitions about mythological figures and saints that had no bearing on Catholic history or teachings. Rather, he argued, they were the result of southerners' ignorance. For instance, Gramsci criticised the well-known devotion to a Donna Bisodia at the time, an entirely fictitious personage believed to be either a prostitute and/or St Peter's mother. This devotion emerged from a misunderstanding of the Latin phrase *dona nobis hodie* (give us this day [our daily bread]) in the Latin version of the Lord's Prayer (*Pater Noster* / 'Our Father').

When I went down to see my family around Pietrelcina, one aunt took me aside and said: 'Your son is so cute! But I think people are jealous; you need to protect him. I know we believe in God and Jesus and everything, but just to make sure, I think we should do the anti-evil eye.' The evil eye (*malocchio*) is the famously Mediterranean form of witchcraft in the sense described by E.E. Evans-Pritchard (1976 [1936]) in his ethnography of the Azande in Sudan: a natural ability, rather than supernatural occurrence, people who possess this type of power can curse others with misfortune (often unwittingly) if they express jealousy (*invidia*) towards them (Dundes, 1992). These people are *jettatori* ('curse-throwers') (De Martino, 1960: 9), but because they are anonymous (sometimes they themselves do not know they have the power), it is never truly certain to whom an unexplained bout of misfortune can be attributed. As Evans-Pritchard writes, this can keep individuals in a community civil to one another, for they never know if those they might offend can 'throw' the witchcraft at them. Since in Italy the *malocchio* is most often cast through *invidia*, it ideally serves the moral purpose of tempering one's envy. Consequently, in the Mediterranean many wear amulets to ward off the evil eye; in Greece, Turkey and Iran these are blue and white glass eyes and can be found everywhere – on buses and taxis, in the home or on the person – but in Italy they take the shape of a horn (Italian-Americans wear these horns around their neck as a sign of their ethnic identity, and often think it is a hot pepper or a symbol of fertility like a cornucopia or even sperm). But nobody is immune; even Padre Pio, as a child, was taken to a folk healer to cure what his parents thought could be an evil eye curse (Alimenti, 1984: 15). This makes sense, for while belief in the *malocchio* is a southern Italian – and pan-Mediterranean – phenomenon, Pietrelcina is only a few kilometres away from the provincial capital, Benevento, which is likened in Italy to Salem, Massachusetts (USA) for the witch trials that took place there during the Inquisition. It is said that witches (called *janare* in the dialect, a possible derivation from the name of the Roman goddess Diana) would meet to dance around an oak tree just outside the walls of Benevento (Oliva, 2015; Stamegna, 2017) – although today the spot is a gas station – and Benevento's most famous drink is a saffron-infused liquor called Strega ('witch' in Italian), complete with an image of that ritual on its label. I've had many informants tell me, 'You know there are still witches in Pietrelcina, right?'. For our part, one of my cousins was having a problem conceiving, and her mother seemed a little jealous of my baby; just to be sure, then, we should do it.

The incantation to ward off the evil eye is shrouded in secrecy, traditionally passed down on New Year's Day from mother to eldest daughter. As the second daughter, my mother does not know how to do it. But although I am not in the proper line or of the right sex and my wife is not related to my aunt (nor Italian), my aunt let me watch the secret rite as my son's father. While I do not want to divulge the specifics of the ritual, it involves utilising a piece of clothing from my son (so that the contagious magic will rub off on him), water and oil (symbols in Catholicism as well as pre-Christian cults), and an incantation that calls on Jesus and local Catholic saints who predate Padre Pio. Observing the rite, it was clear that this was a syncretic practice, one that fluidly and creatively integrates Christian theology with folk healing; my aunt's discursive justification ('we believe in Jesus, but just to be sure …') also revealed the cognitive dissonance (Festinger, 1957) necessary in popular religiosity: these practitioners are not ignorant of 'proper' theology, nor about the seemingly 'backwardness' that such practices are considered in the modern era, but rather have a nested system of healing practices, just as Americans do, for solving well-being issues. If a doctor doesn't work and the priest doesn't work, then a folk healer or witch doctor will be approached. Indeed, it became clear in collecting stories about Padre Pio's miraculous healing abilities that he was himself treated at the time as a sort of folk healer by Italians, one who was seen to mix his closeness with the supernatural with a very natural ability to effect changes in one's health and well-being for devotees through extraordinary material powers.

The Baptism: Achieved Kinship, Tourism and the Performance of Religion

While my foray into evil eye incantations to protect my son provided a useful insight into the lived experiences and mindset of popular religious practitioners within the devotional cult to Padre Pio, my son's baptism was an entirely different affair. Thanks to my in-laws, who during their time as employees in the US Embassy made friends with a young Italo-Argentinian priest who would become a prominent Cardinal in the Vatican, we were able to arrange his baptism in St Peter's Basilica, the heart of the Catholic Church. Here, the power of religious authority was on full display. We had agreed to have our son baptised in the traditional, marble font, which the Church had converted from the tomb of pagan Roman emperor Hadrian. It sits in a Baroque chapel immediately to the left of the doorway when one walks into the basilica, in full view of the public (although it is gated). The Cardinal was to personally preside, assisted by my friend from Pesaro.

The baptism was a grand affair. Its position in the Basilica – which sees roughly 40,000 visitors daily – rendered it semi-private at most. Glancing through the ornate gated bars that separated the baptistry from the nave (the central aisles of the church), I could see a crowd of tourists watching and taking pictures throughout the ceremony; the Cardinal with his mitre and crosier (hat and staff) seemed to be captivating to the visitors. Later the next morning, when we were back in the hotel, some of these tourists ran into us, and were star-struck: 'You're those actors that were doing the baptism yesterday in St Peters, right? How many times a day to you do that performance for the visitors?', they asked. We explained that we were not actors, but rather that St Peter's is a living church – it was a real rite, not a performance for the tourists as they suspected!

As a practising Catholic studying pilgrimage and tourism, the experience was illuminating. I had assumed that spectators would understand that such ceremonies were real. Rather, it emphasised the power of tourist imaginaries (Salazar & Graburn, 2014) to shape perceptions of authenticity, and what constituted the touristic frontstage and backstage (MacCannell, 1976). It also showed the power of 'heritagisation' to sometimes ossify or museumify a site (see Di Giovine & Garcia-Fuentes, 2016; Isnert & Cerezales, 2020), particularly when a site suffers what is today called 'overtourism' – unbridled visitation that puts stresses on the local community. It is an issue that I subsequently tackled in my research, particularly as managers at Pio's shrine endeavour to convert their site from a place of popular devotion to one of religious heritage (Di Giovine, 2015b). Both Pietrelcina and Pio's shrine of San Giovanni Rotondo endeavour to diversify their offerings, to draw more than simply religious pilgrims; this is a means of ensuring viability if (or when) Pio devotion is eclipsed by other more contemporary saints in the future. The challenge, of course, is to maintain the living nature of faith.

Conclusion: The Return Home

Using autoethnographic vignettes, this chapter reflected on the myriad ways in which one's kinship and fatherhood in the field ultimately shape anthropological research. While other forms of social scientific research – particularly those of a quantitative nature – may endeavour to resist or defer subjectivity, ethnography is predicated on bodily engagement with the field site and the social life contained therein. One's own status – as a kinsman or outsider, father or son, citizen or foreigner – will not only shape the way in which the researcher perceives the social world studied, but veritably how that social world comes alive for him. As this chapter has shown, the biological and social nature of being a father and kinsman not only humanises the foreign ethnographer-Other – providing a sense that 'he' is just like 'us' in some core way, which builds trust and ensures more candid and 'human' discussions – but also opens doors for different sorts of experiences and conversations about life and living, growth and education, culture and identity.

And it provides an entrée into other worlds, worlds unseen and perhaps un-sought-after during the pre-field planning stages of the research – but worlds that deeply surround the research topic. Using the self 'as the instrument of knowing,' ethnography, as Sherry Ortner states, is a method 'in which the whole self physically and in other ways enters the space of the world the researcher seeks to understand' (Ortner, 1995: 173). The deeper one is enmeshed in the minutiae of local daily life, viscerally and bodily experiencing it, the deeper and more robust is one's cultural understanding. The 'emplaced body and lived experience (of the researcher) can yield deeper levels of understanding and insight through engaged practices on different levels of immersion', states Ann David (2013: 45). Although I had to perform my affective kinship, the kin group I was born into – and the achieved kinship group into which I married – served as unique and privileged networks that set the stage for a breadth of experiences that was largely unattainable on my own. Furthermore, the decisions I made to ensure the well-being of my wife and son pushed me to perform my citizenship, religious affiliation and political connections – which illuminated largely latent racial, political and economic dynamics in Italian society. These also allowed me to face the privileges of my status and mobility in ways I had not before.

Such statuses also shape the subsequent analysis and its presentation, as feminist and queer anthropology show. As Evans-Pritchard states, such a visceral form of inquiry

> derives not merely from intellectual impressions of native life but from its impact on the entire personality, on the observer as a total human being. … The work of the anthropologist is not photographic. He has to decide what is significant in what he observes and by his subsequent relation of his experiences to bring what is significant into relief. (Evans-Pritchard, 1951: 82)

Indeed, relationship forging is an important component of anthropological research, an arena for obtaining seemingly authentic information. Malcolm Crick (1995: 216) points out that all anthropologists typically seek to endear themselves to their research subjects, and represent them especially in subsequent publications as their friends or quasi-kin. They do this 'for quite explicit aims; they deliberately forge relationships in order to obtain data; they have their eyes on publications even before they embark on their field work'. But relationship building and the creation of fictive kinship with their research subjects is also implicated in processes of anthropological self-fashioning, of presenting a version of oneself as a quintessentially immersed ethnographer who was able to build trust and 'go native'. Yet

> anthropologists are also conscious of the fragility of their position, living with the 'working fiction' (Geertz, 1968: 151–154) that they share a world of understanding with their informants, but aware that at any time that world may be shattered and their situation disintegrate into one lacking trust, lacking understanding, in fact, exploded into two mutually non-communicating worlds. (Crick, 1995: 216)

What I argue here is that this impetus – which we are trained to do – was enhanced through my affective and achieved kinship status. Sometimes I situationally deployed it, but at other times it simply opened doors in serendipitous ways.

Becoming a father in the field was an extraordinary experience, just as it would be when my second son was born back at home in the United States. But the experience of fatherhood in Italy was also integral, and inexorably linked, to my research and its outcome. It thrust me into situations that were unexpected and unplanned, yet that ultimately were revelatory of the broader cultural context in which my research was situated. Furthermore, creating a family and attempting to care for my wife and child forced me to acculturate in ways much more fundamental than the typical expat ethnographer would be willing to do. And it brought to the fore the imponderabilia of social life – in both Italy and back at home – that had been naturalised and implicit to me. Indeed, my son shaped my research, and the research shaped me. I was fundamentally changed by my fieldwork, which was fundamentally conditioned by my son's birth. As Georges Condominas (1973: 2) writes, 'the most important moment of our professional life remains fieldwork: at the same time our laboratory and our *rite de passage*, the field transforms each of us into true anthropologists'.

Eight months after our son was born, we made our preparations to leave Italy and return home to Chicago so I could write up my research. This had been planned from the beginning, and while we looked forward to returning to our family and friends in the United States – our new addition in tow – I also looked at our departure with regret. For a 'hyphenated'

American (Minh-ha, 1991: 159) such as myself, who straddles Italian and American culture, becoming a father in Italy was a way of returning to my roots, of not only embracing my Italianness, but playing it out. But leaving Italy also meant that I would take my son out of that context: he would surely enculturate as an American; his Italianness would fade away. I shared this regret with several of my Italian friends and family, but they all made recourse once again to *terroir*: he is Italian because he was born here, he developed here. His time in the womb, his first breath, first sounds, first taste of food were all Italian. He is inexorably tied to the land, and the land is tied to him – and thus to me. And so it came full circle: born from an Italian in America, I had returned to create a new generation back in the homeland. In that, he fulfilled my desire for rootedness; he completed my identity, and strengthened our connectedness with our kin.

He was my Chianti.

References

Alimenti, D. (1984) *Padre Pio*. Bergamo: Velar.
Banfield, E. (1958) *The Moral Basis of a Backwards Society*. New York: Free Press.
Barth, F. (1956) *Ethnic Groups and Boundaries*. Boston, MA: Little, Brown.
Boellstorff, T. and Howe, C. (2015) Queer futures. Editor's forum: Theorizing the contemporary. *Cultural Anthropology*, 21 July. See https://culanth.org/fieldsights/series/queer-futures (accessed 15 January 2020).
Carroll, J. (1992) *Madonnas that Maim: Popular Catholicism in Italy since the Fifteenth Century*. Baltimore, MD: Johns Hopkins University Press.
Cavaioli, F. (2008) Patterns of Italian immigration to the United States. *Catholic Social Science Review* 13, 213–229.
Clifford, J. and Marcus, G. (1986) *Writing Culture*. Berkeley, CA: University of California Press.
Cole, J. (1997) *The New Racism in Europe: A Sicilian Ethnography*. Cambridge: Cambridge University Press.
Collins, P. and Gallinat, A. (2010) *The Ethnographic Self as Resource*. Oxford: Berhghan.
Comaroff, J. and Comaroff, J. (1992) *Ethnography and the historical imagination*. Boulder, CO: Westview Press.
Condominas, G. (1973) Distinguished lecture 1972: Ethics and comfort – an ethnographer's view of his profession. *Annual Report*. Washington, DC: American Anthropological Association.
Corriere della Sera (2018) Gli anni passano. La sfiducia degli italiani, no. *Corriere della Sera*, 29 March. See https://www.corriere.it/sette/18_marzo_29/fiduciaitalia-654e55e0-3119-11e8-b98c-6b7fd54f26e4.shtml (accessed 14 January 2020).
Crick, M. (1995) The anthropologist as tourist: An identity in question. In M.-F. Lanfant, J.B. Allcock and E.M. Bruner (eds) *International Tourism: Identity and Change* (pp. 205–223). Thousand Oaks, CA: Sage.
David, A. (2013) Ways of moving and thinking: The emplaced body as a tool for ethnographic research. In P. Harrop and D. Njaradi (eds) *Performance and Ethnography: Dance, Drama, Music* (pp. 45–66). Newcastle upon Tyne: Cambridge Scholars.
Davis, J. (1998) Casting off the 'southern problem': Or the particularities of the south reconsidered. In J. Schneider (ed.) *Italy's Southern Question* (pp. 205–224). Oxford: Berg.

De Martino, E. (1960) *Sud e magia*. Milano: Feltrinelli.
Di Giovine, M.A. (2009) *The Heritage-scape: UNESCO, World Heritage, and Tourism*. Lanham, MD: Lexington Books.
Di Giovine, M.A. (2010) La Vigilia Italo-Americana. *Food and Foodways* 18, 181–208.
Di Giovine, M.A. (2012a) Making saints, (re-)making towns: Pilgrimage and devotion to St. Padre Pio of Pietrelcina. Unpublished dissertation, University of Chicago.
Di Giovine, M.A. (2012b) Padre Pio for sale: Souvenirs, relics or identity markers? *International Journal of Tourism Anthropology* 2 (2), 108–127.
Di Giovine, M.A. (2012c) Passionate movements: Emotional and social dynamics of Padre Pio pilgrims. In D. Picard and M. Robinson (eds) *Emotion in Motion: Tourism, Affect and Transformation* (pp. 117–136). Farnham: Ashgate.
Di Giovine, M.A. (2015a) The everyday as extraordinary: Revitalization, religion, and the elevation of Cucina Casareccia to heritage cuisine in Pietrelcina, Italy. In R. Brulotte and M.A. Di Giovine (eds) *Edible Identities* (pp. 77–92). Farnham: Ashgate.
Di Giovine, M.A. (2015b) When popular religion becomes elite heritage: Tensions and transformations at the shrine of St. Padre Pio of Pietrelcina. In H. Silverman and M. Robinson (eds) *Encounters with Popular Pasts: Cultural Heritage and Popular Culture* (pp. 31–47). New York: Springer.
Di Giovine, M.A. and Garcia-Fuentes, J.-M. (2016) Sites of pilgrimage, sites of heritage: An exploratory introduction. *International Journal of Tourism Anthropology* 5, 1–23.
Dundes, A. (1992) *The Evil Eye: A Casebook*. Madison, WI: University of Wisconsin Press.
Durkheim, É. (1982 [1895]) The Rules of Sociological Method (W.D. Halls, trans.). New York: Free Press.
Engelke, M. (2004) 'The endless conversation': Fieldwork, writing and the marriage of Victor and Edith Turner. In R. Handler (ed.) *Significant Others: Interpersonal and Professional Commitments in Anthropology*. Madison, WI: University of Wisconsin Press.
Evans-Pritchard, E.E. (1951) *Social Anthropology*. London: Cohen & West.
Evans-Pritchard, E.E. (1976 [1936]) *Witchcraft, Oracles, and Magic among the Azande*. Oxford: Oxford University Press.
Femminella, F. (1961) The Impact of Italian migration and American Catholicism. *American Catholic Sociological Review* 22 (3), 223–241.
Festinger, L. (1957) *A Theory of Cognitive Dissonance*. Stanford, CA: Stanford University Press.
Frazer, Sir J. (1958 [1922]) *The Golden Bough: A Study in Magic and Religion* (abridged edn). New York: Macmillan.
Friedrich, P. (1977) *Agrarian Revolt in a Mexican Village*. Chicago, IL: University of Chicago Press.
Geertz, C. (1968) Thinking as a moral act: Ethical dimensions of anthropological fieldwork in the New States. *Antioch Review* 28 (2), 139–158.
Geertz, C. (1973) *The Interpretation of Cultures*. New York: Basic Books.
Gramsci, A. (1926) *La Questione Meridionale* (F. De Felice and V. Parlato, eds). Rome: Editori Riuniti.
Gramsci, A. (1996) *Lettere Dal Carcere, 1926–1937, Vol. 2* (A. Santucci, ed.). Palermo: Sellerio Editore.
Handler, R. (2000) *Excluded Ancestors, Inventible Traditions: Essays toward a More Inclusive History of Anthropology*. Madison, WI: University of Wisconsin Press.
Handler, R. (2010) *Significant Others: Interpersonal and Professional Commitments in Anthropology*. Madison, WI: Unviversity of Wisconsin Press.
Isnert, C. and Cerezales, N. (2020) *The Religious Heritage Complex: Legacy, Conservation and Christianity*. London: Bloomsbury.
Laurino, M. (2000) *Were You Always Italian? Ancestors and Other Icons of Italian America*. New York: W.W. Norton.

Lutz, C. (1990) The erasure of women in sociocultural anthropology. *American Ethnologist* 17 (4), 611–627.
MacCannell, D. (1976) *The Tourist: A New Theory of the Leisure Class*. New York: Schocken Books.
Malinowski, B. (1984 [1922]) *Argonauts of the Western Pacific*. Long Grove, IL: Waveland Press.
Minh-ha, T.T. (1991) *When the Moon Waxes Red: Representation, Gender and Cultural Politics*. New York: Routledge.
Ministero dell'Interno (2020) *Voto degli Italiani all'Estero*. See https://www.interno.gov.it/it/temi/elezioni-e-referendum/voto-italiani-allestero (accessed 13 January 2020).
Moore, H. (1988) *Feminism and Anthropology*. Cambridge: Polity Press.
Murphy, Y. and Murphy, R. (1974) *Women of the Forest*. New York: Columbia University Press.
Myerhoff, B.G. (1978) *Number Our Days: A Triumph of Continuity and Culture among Jewish Old People in an Urban Ghetto*. New York: Simon & Schuster.
Newton, E. (1993) My best informant's dress: The erotic equation in fieldwork. *Cultural Anthropology* 8 (1), 3–23.
Oliva, A. (2015) *Le Streghe di Benevento: La Leggenda della 'Superstitiosa Noce'*. Vasto: Caravaggio Editrice.
Ortner, S. (1995) Resistance and the problem of ethnographic refusal. *Comparative Studies in Society and History* 37 (1), 173–193.
Ortner, S. (1974) Is female to male as nature is to culture? In M. Rosaldo and L. Lamphere (eds) *Women, Culture, and Society*. Stanford: Stanford University Press.
Salazar, N. and Graburn, N. (2014) *Tourism Imaginaries: Anthropological Approaches*. Oxford: Berghahn.
Schneider, J. (ed.) (1998) *Italy's 'Southern Question': Orientalism in One Country*. Oxford: Berg.
Stamegna, M. (2017) *Miti e Leggende del Centro-Sud Italia*. Gaeta: AliRibelli Edizioni.
Stoller, P. (1999) *The Taste of Ethnographic Things: The Senses in Anthropology*. Philadelphia, PA: University of Pennsylvania Press.
Strathern, M. (1988) *The Gender of the Gift*. Berkeley, CA: University of California Press.
Strathern, M. (1999) *Property, Substance, and Effect: Anthropological Essays on Persons and Things*. London and New Brunswick, NJ: Athlone Press.
Trubek, A.B. (2008) *Taste of Place: A Cultural Journey into Terroir*. Berkeley, CA: University of California Press.
Turner, V. and Bruner, E. (1986) *The Anthropology of Experience*. Urbana, IL: University of Illinois Press.
Turner, V. and Turner, E. (1978) *Image and Pilgrimage in Christian Culture*. New York: Columbia University Press.
Whiting, B. and Whiting, J. (1975) *Children of Six Cultures: A Psychocultural Analysis*. Cambridge, MA: Harvard University Press.
Wilkinson, T. (2008) Padre Pio exhumed for a second life. *Los Angeles Times*, 25 April. See https://www.latimes.com/archives/la-xpm-2008-apr-25-fg-padre25-story.html (accessed 27 October 2020).

Masculinities in Tourism Research: Implications and Conclusions

Joseph M. Cheer, Heike A. Schänzel and Brooke A. Porter

> Fast-forward to today and there are much more nuanced and realistic models of masculinity attuned to the realities of 21st century living. No longer is there a sense that men have to be strong in every sense of the word.
>
> Wired, 2018: para. 3

That the gender equality imperative and its attendant gender wars are long established, signals that the overarching themes around gender and its many contentions, contradictions and perpetuations in the everyday are undeniable. The continuing ascendency of women in a world that is arguably, on balance, still very much a 'man's world' is due to the perseverance of women in their battle for equality, empowerment and social justice. This, without doubt, disrupts the balance. Thus, like any transition, the trajectory is intense, bumpy and replete with anxiety and discomfort for some, and a source of hope and vindication for others. The bumpiness is alluded to in the opening quotation with the consequence that what being a man means is continually being reshaped amid the ongoing resetting of gender relations. Whether this is considered indulgent naval gazing by the intelligentsia and educated middle class, or timely and necessary in contemporary societies, is up for contention. Notwithstanding, gender relations are clearly pivoting away from hackneyed conceptualisations of masculinity, albeit not without a rear-guard action from constituents that remain rooted in the traditional past.

Nevertheless, the gender empowerment imperative is writ large across the globe, and the best exemplifications of this are the United Nations Sustainable Development Goals (SDGs), where the underpinning aims are to give women and girls the same rights their male counterparts already enjoy. Whether such goals are lofty and unrealistic, or whether this is long overdue, is framed by social constructions and norms that generalise all manner of things that have long been related to gender. Undoing what has been set in stone will take some focused effort. Such sentiments permeate all walks of life and continue to manifest in the struggles of women's movements, impacting male counterparts who are supportive and sympathetic, as well as those who are belligerent, resentful and firmly set in their ways. Nevertheless, while change processes are in play, when it comes to conceptualisations of contemporary masculinity, this disruption or revision of the status quo does not go unchallenged or unequivocally embraced, nor is it received willingly, as renowned masculinities scholar Michael Kimmel's reflections suggest:

> My father's world was like Don Draper's: Everybody knew their place. The men smoked in elevators and drank hard liquor during the weekday. And I grew up thinking that my world would look like that, and it looks nothing like that. But my son has no such expectations, and he knows it and he's fine with that. Young men say being nurturing, caring, and being a great dad is what being a man is about. (Krasny, 2015: para. 5)

In this volume, we have taken the wider and overarching gender empowerment endeavour and aligned it to academic research practices. We pose several questions but, most importantly, we set out to establish how the performance of masculinities plays out in the field by asking researchers, all of whom are male, how gender has influenced and shaped their research practice. While we acknowledge the political and moral responsibilities associated with disrupting gender relations, the emphasis remains on conveying the experiences of contributors and their articulations of how they have dealt with the overarching contentions that they encountered. In response, contributors have engaged in deep reflexive thought, reversing the mirror back on themselves and prompting critical self-examination. In trying to get answers, we have asked for somewhat unconventional approaches to scholarly discourse, not bound by empirical data and stringent methodological cross-examinations but, instead, by using autoethnography. Autoethnography calls for deep reflection – at times confronting and uncomfortable for some, while for others a cause for relief and expressions of what has hitherto been suppressed, under acknowledged or conveniently ignored.

That the call to arms for almost all contributors resulted in the question of what it means to be a man in the present day is revealing. Is it that men and boys are caught in a bind underlined by not knowing how they should be as men and boys? Is it also possible that this might be related to the transition taking place, where once it might have been unthinkable to pose such a question, but now the gender empowerment imperative has shaken things up and the dust has yet to settle? Most importantly, does this signal that for men and boys the status quo sits on shifting ground, where role models who were once invincible are now being stripped of the veneer that shielded them, leaving them naked and exposed – warts and all, so to speak – having to learn how to navigate towards a full expression of their gender?

In a sense, this volume depicts contemporary life, where ongoing negotiations take place and the happy medium where optimum gender relations might be continues to remain elusive, or at the very best fleetingly showing its face. The idea of gender itself as a binary between one and the other has been summarily dismantled, and the push by social progressives has been towards acknowledging a gender spectrum, as it were, and not a neat duality for all to fit into. For academic researchers, this is the reality of the backdrop against which they work and the spaces within which their respondents and informants reside. However, as exemplified by extant research in tourism-related studies specifically, contemplation and queries into gender have almost exclusively come from female researchers (e.g. Cole, 2018; Figueroa-Domecq & Segovia-Perez, 2020; Porter & Schänzel, 2018). Moreover, much of this is focused on the quest for equality, empowerment and inclusiveness in tourism – appeals for fairer outcomes. Consequently, integrating considerations of gender and its various assemblages into research practice is imperative. Yet, for many contributors, thinking about gender and the application of their masculinities was unprecedented. Furthermore, that the maleness of contributors could influence their research practice and somehow potentially shape their fieldwork outcomes was rarely, if at all, considered in fieldwork planning and research design phases. Most researchers articulated that they simply 'got on with the job', giving short shrift to understanding how their masculinities might moderate or determine the fecundity of fieldwork undertakings. This speaks to the necessity for a rethinking of the way masculinities are conceived of and acted out in academic practice. However, it seems that any reconfigurations or system redrawing will more likely be achieved through external pressure. As Donaldson (1993) implores:

> if the gender system has an independence of structure, movement, and determinations, then we should be able to identify counter-hegemonic forces within it; if these are not identifiable, then we must question the autonomy of the gender system and the existence of hegemonic masculinity as central and specific to it. (Donaldson, 1993: 644)

As surprising as this might seem, this is exactly why this volume has emerged: firstly, from uncovering the vitalities of femininities in fieldwork practice and realising how engaged female researchers were with their gender and its many aspects, especially its practical limitations; and secondly, through realising how feminist theory and wider thought underpin the work done by female scholars within tourism research. This stood in stark contrast to our observations of male scholars who ploughed headlong into research focused on overcoming technical, theoretical and practical concerns, with rarely an acknowledgement of how their masculinities might potentially influence their practice. In a sense, hegemonic masculinities are conceived of as not being in need of consideration. This is a reminder of Donaldson's (1993: 645) retort that 'Heterosexuality and homophobia are the bedrock of hegemonic masculinity and any understanding of its nature and meaning is predicated on the feminist insight that in general the relationship of men to women is oppressive'. Thus it is no surprise that, in this volume, the exception to this lack of gender consideration by men appears to be non-heterosexual males, probably by virtue of the precarity that they experience in the reception of and attitudes to their sexualities – often in negative and oppressive manifestations. This has meant that, for this cohort of non-heterosexual male researchers, their demonstration of masculinity in research practice has been strongly configured in relation to their experienced struggles for acceptance and normalisation.

Emergent Queries

- **What does it mean to be a man?**
  This is the overarching theme emerging from our contributors and is related to the all-consuming effort of 'proving' to be a man anywhere in the world, which in fact can seem a ridiculous and anachronistic concept. Women would never entertain the concept of 'proving' that they are a woman (Biddulph, 2013). For female researchers (see Porter & Schänzel, 2018), the struggles were largely in the world around them, while for male researchers the struggles were often from the inside. Why does this matter? Because it gets us to the heart of the broader gender issue in that identifying the diverse types of 'manliness' can help us understand areas of concern and ways forward beyond a need to have to define what it means to be a man and the need for any proof.
- **Why the resistance to emotional entanglements?**
  For the most part, authors reflected a combination of ambivalence and uncertainty when it came to emotional entanglements, and these eventually presented as resistance. This is reminiscent of Sideris's (2004: 46) contention that 'such appeals to "the tradition" represent a resistance to change, a retreat to a fixed set of principles by which to organise behaviours and relationships that admit neither interrogation nor alternative customs'.

  That (generally heterosexual) men shy away from emotional entanglement might be changing due to a shift in generational attitudes. While generational attitudes may allow for (generally heterosexual) men to face emotional entanglements, if male reluctance to becoming emotionally involved continues as an abiding condition, its ability to enforce limitations on male researchers building respondent relations and rapport to good effect deserves some circumspection. The contributions presented in this volume suggest a lingering rigidity and steadfastness in the face of transitioning.

- **Why are masculinities still taken for granted?**
  The predecessor to this volume, *Femininities in the Field* (Porter & Schänzel, 2018), was completed well ahead of schedule, which differed markedly from the experience of producing this male-authored companion work. While not the rule and possibly a generalisation on behalf of the editorial team, timeliness and respect for deadlines (pre Covid-19) in this case was inferior with a pattern of justifying excuses from authors submitting their chapters well beyond stated deadlines. This raises the question as to whether women are more prone to playing by the rules than men. Further related questions raised include: How is the dominance of masculinities shaping research insights and knowledge production processes? How would including more diverse gendered voices in the research process impact our insights? Perversely, this is also related to access to study sites that may exclude researchers on the basis of their gender.
- **How is access to research sites and participants gendered?**
  In this book, several contributors lamented that, by virtue of being male, their access to certain research sites and contexts was constrained. This was often decried and considered unfair and needless. However, these interpretations under-acknowledge the extent to which female researchers experience impeded access to study sites in their research practice, or that, unlike men, in some situations women may need safe spaces, as seen in the development of female-only facilities in hotels and on public transport. Pertinent questions are raised: Where do we draw the line in allowing males access to women-only safe spaces? How do we maintain respect for these inner sanctuaries of women? Questions of access inhabit a duality and are also related to taking masculinity privileges for granted.
- **How do we include a spectrum of masculinities?**
  In relation to vulnerable masculinities, the experiences and challenges of non-heteronormative contributors as well as those from ethnically diverse backgrounds in this volume are testament that developing competencies to successfully negotiate diverse masculinities is vital.
- **How do we address toxic masculinities?**
  Although not discounting its existence, the occurrence or issue of toxic femininity was absent in the first project (see Porter & Schänzel, 2018). We hear a considerable amount about toxic masculinity but far less about toxic femininity. Devon Price, a self-referenced nonbinary social psychologist and writer, places the origin of toxic genders in our strict adherence to the gender binary.

  > Focusing only on the harm done by men – and the insecurities harboured by men – ignores the broader, systematic nature of the beast. The problem was never just masculinity. It was, and is, inflexible gender roles for men and women alike. (Price, 2019: para. 9)

Specifically, Price considers toxic masculinity as 'a warped, unrelenting cultural relationship to masculinity' (Price, 2019: para. 4). The trauma that such a relationship creates will take a long time to resolve but needs wider acknowledgement in order to find such resolution. During the first rounds of edits, we found statements or descriptions from authors that would have likely offended many females. While we truly doubt this was the intention of any of the contributors, it leads to the questions: How do we ensure that safety is associated with self-expression? How is gendered privilege used, either knowingly or unknowingly? These questions are further related to seeking more emotional entanglements, something that should be considered desirable.

- **How should we include the consideration of masculinities in research practice?**
We should actively and immediately consider adding a gendered dimension to research training, reflecting on both femininities and masculinities to expand and deepen tourism research with a true gender perspective. For research that has already occurred, there is the opportunity to include it in the writing. Some have already taken notice. For example, Mach (2019), in a general study on surf tourism, explicitly noted his gendered limitations, among others:

> There are identifiable limitations with the findings in this article. The informants were mostly comprised of members of one of the first cohorts to pass through the surf lesson program. While this enhanced the ability to gather participant reflection on participation over time, there is limited representation from other cohorts who may have had different experiences and interpretations of the program. In addition, while gender emerged as a critical discussion topic, I recognize that being a male researcher may have limited the depth of information that female informants were willing to share and that my own gender socialization may have skewed my interpretation of the data collected. I tried to correct for this by providing representative quotes directly from study participants. (Mach, 2019: 444)

To begin this journey, we call for anyone engaged in research to incorporate reflexivity, even if only a single paragraph, into their writing. We recognise that, for some, the incorporation of masculinities (and gender, in general) will be more nuanced. We have previously noted the need for the male supervisors of female students to pay attention to gender (Porter & Schänzel, 2018) and now we widen this call for care to the supervisors of non-heteronormative and other non-gender conforming students. Doing so will also allow for critical retrospection of possible power imbalances and other limitations. In the end, our learnings and insights are not about gender itself; rather, they encompass the notion that we all possess the same humanity. This oneness of selves requires constant deep reflection to counter any possible imbalances due to gender, ethnicity or otherwise.

Research Implications: The Path Forward

This volume highlights that, hitherto, little attention, in both formal and informal research contexts, has been given to considerations of masculinity and how it shapes and determines the extent to which male researchers perform and are received while practising research. Myriad research implications must follow, both practical and theoretical. One upshot might be the links between masculinities and work in a post-feminist context. As Rumens (2017: 247) opines, post-feminism is a sensibility that acknowledges feminism and 'is seen to confront the challenge of theorizing difference by abandoning the binary thinking of second-wave feminist theory and focusing on plurality, fluidity and hybridism'. Post-feminism also acknowledges both the small and the seismic shifts that have taken place as well as everything in between.

Perhaps, in performing masculinities in the field during the process of formal scholarly endeavour, post-feminism offers researchers a theoretical framework from which to construct their conceptualisations of participants (particularly female) in a manner that promotes constructive and mindful encounters. Post-feminism might also present male scholars with a mirror to reveal their own tightly held constructions of gender and the many notions and generalisations that underpin it. This is reflected very broadly in contributions to this book where the reflexive gaze is mobilised and all at once authors have been corralled into questioning the application of their craft with a firm eye fixed on the manifestations of their masculinities. Indeed, as Rumens (2017) highlights, 'postfeminist discourses of female empowerment, individualism and choice must be understood as a technology of power' and its very application can engender discourses and interactions that pursue more fecund arrangements for male researchers in the act of playing out masculinities and the assemblages attached to them. Moreover, there is little doubt that the moral and ethical responsibilities of male researchers in helping address gender disparities remain – this is what we have aimed to emphasise in the questions foregrounded earlier.

Prescribing a fool-proof method for the way forward would be presumptuous given that the gender landscape is waxing and waning with variegations across the multitude of global contexts within which researchers work. The calibration of gender relations and the extent to which conceptions of masculinities stay fixed (or not) must surely remain fluid, requiring scholars to assess and reassess their fieldwork practice repeatedly. Notwithstanding, if the nuanced sensibilities of fieldworkers can discombobulate constructions of gender relations, it might well be the conduit to more productive fieldwork outcomes. After all, as prominent masculinities scholar Michael Kimmel argues, 'although we're more gender equal than ever before, we have a long way to go' (Krasny, 2015: para. 9). We end with a call to consider how understanding masculinities and the reflective research practices of male-identifying researchers and scholars can offer part of the solution for creating a more gender equitable world. Thus, we think scholarship on masculinity, and our own research endeavours, can make a difference in building up a shared sense towards a more inclusive humanity.

References

Biddulph, S. (2013) *The New Manhood: The 20th Anniversary Edition*. Sydney: Simon & Schuster.

Cole, S. (ed.) (2018) *Gender Equality and Tourism: Beyond Empowerment*. Wallingford: CABI.

Donaldson, M. (1993) What is hegemonic masculinity? *Theory and Society* 22 (5), 643–657.

Figueroa-Domecq, C. and Segovia-Perez, M. (2020) Application of a gender perspective in tourism research: A theoretical and practical approach. *Journal of Tourism Analysis/Revista de Análisis Turístico*.

Krasny, J. (2015) What does it mean to be a man? This professor might have the answer: A conversation with Michael Kimmel, the founder and director of the Center for the Study of Men and Masculinities. *Esquire*, 9 September. See https://www.esquire.com/news-politics/news/a37762/what-does-it-mean-to-be-a-man/ (accessed 10 January 2019).

Mach, L. (2019) Surf-for-development: An exploration of program recipient perspectives in Lobitos, Peru. *Journal of Sport and Social Issues* 43 (6), 438–461.

Porter, B.A. and Schänzel, H.A. (eds) (2018) *Femininities in the Field: Tourism and Transdisciplinary Research*. Bristol: Channel View Publications.

Price, D. (2019) Toxic femininity holds all of us back: Men and women alike perpetuate and suffer from it every day. *Human Parts*, 31 December. See https://humanparts.medium.com/toxic-femininity-is-a-thing-too-513088c6fcb3 (accessed 7 January 2020).

Rumens, N. (2017) Postfeminism, men, masculinities and work: A research agenda for gender and organization studies scholars. *Gender, Work & Organization* 24 (3), 245–259.

Sideris, T. (2004) 'You have to change and you don't know how!': Contesting what it means to be a man in a rural area of South Africa. *African Studies* 63 (1), 29–49.

Wired (2018) What does 'being a man' mean? *Wired Magazine*, 2 March. See https://www.wired.co.uk/article/what-does-being-a-man-mean-anyway (accessed 28 December 2019).

Index

absence(s) 120, 123, 172, 191, 192, 194, 196
accept (ance), (ed) 2, 15, 20, 31, 34, 46, 49, 52, 53, 54, 61, 74, 78, 83, 113, 118, 126, 131, 140, 146, 153, 154, 157, 159, 161, 162, 164, 165, 167, 173, 212, 231
access (ing) xiv, 6, 7, 35, 49, 51, 52, 53, 60, 61, 62, 63, 65, 80, 114, 11, 117, 120, 124, 134, 139, 142, 144, 170, 171, 172, 173, 176, 182, 198, 232
active interviewing method 85
adventure(s) viii, xiv, 6, 13, 46, 48, 49, 50, 51, 52, 53, 58, 132
advisor 117
Africa(n) 6, 56, 57, 58, 60, 62, 153, 154, 155, 156, 157, 158, 159, 160, 161, 162, 163, 165, 166, 167, 214
ambivalence 88, 96, 139, 212, 231
anger 26, 88, 131, 172, 178
anthropology 85, 136, 142, 148, 206, 208, 209, 223
anthropology of experience 209
art world 89, 91, 92
āsana 141
asymmetric(al) 130, 179
autoethnography 4, 19, 20, 21, 22, 23, 25, 103, 123, 124, 125, 207, 229
awkward encounters 85, 86, 88, 89, 92, 95

baptism 222
Benevento 221
Bethlehem 22, 23, 24, 25, 26, 171, 177, 178, 179, 180, 181
bias(es)(ed) xiv, 4, 5, 6, 32, 36, 46, 54, 89, 91, 94, 95, 116, 118, 119, 147, 200, 202
bibliometric study 21
blacklisted 182
blockade(s) 175, 183
bloke-ism 49
Botswana 154, 155, 161, 165

campanilismo 211, 217
Catholic church 222
Catholicism 72, 218, 221
challenges 7, 51, 53, 57, 60, 65, 80, 88, 104, 105, 126, 175, 194, 195, 199, 201, 232
checkpoints 170, 175, 176, 177, 178, 179, 180, 181, 182, 183
childbirth 217, 218
Chinese 91, 106, 114, 116, 117
Chineseness 114, 116
Christian 74, 173, 221
citizenship 214, 215, 216, 217, 223
class xiv, 76, 125, 126, 127, 128, 130, 131, 144, 146, 179, 209, 210, 218
closure 176
colonial 6, 27, 72, 125, 127, 128, 129, 130, 131, 136, 137, 167, 176, 216
colonial project 136
colonisation 18, 19, 72, 157, 158, 160
conference(s) 6, 25, 34, 35, 41, 116, 117, 142, 213
conflict-ridden destination xiv, 182
contested spaces 23
courage 48, 51, 52, 53, 157
cross-cultural xiv, 7, 102, 104, 105, 106, 119, 121, 167
cultural 18, 19, 20, 23, 52, 71, 74, 76, 83, 101, 103, 104, 106, 108, 110, 111, 113, 114, 121, 128, 130, 135, 137, 141, 142, 143, 153, 154, 155, 158, 161, 163, 164, 165, 166, 167, 172, 191, 195, 207, 209, 210, 214, 217, 218, 219, 223, 224, 233
culture(s) xiii, 5, 7, 19, 20, 61, 64, 71, 72, 75, 80, 86, 93, 95, 111, 134, 136, 137, 139, 140, 144, 153, 154, 156, 157, 161, 163, 165, 167, 179, 206, 207, 208, 209, 216, 217, 223, 225
culture shock 165, 166, 193, 195

decolonise yoga 141
Deleuzian constructs 125
denied 18, 20, 183
dependent 15, 114, 167
despair 178, 179, 180
disparagement 49
disruption 198, 199, 201, 229
dissonant xiv, 7, 134, 160, 163, 164, 165, 167
doing research 81, 125, 182

East Jerusalem 173
emasculation 179
emotional writing 6, 17
emotions xiii, 13, 15, 16, 17, 18, 19, 20, 21, 22, 23, 25, 47, 52, 88, 91, 105, 170, 171, 172, 178, 179, 180, 182, 207
epistemic relativism 136
empowerment 34, 43, 111, 112, 113, 228, 229, 230, 234
empty signifier 140
ethics 41, 56, 131
ethnicity 87, 90, 93, 94, 116, 117, 119 , 123, 134, 146, 147, 164, 223
ethnography 85, 88, 89, 90, 91, 92, 93, 96, 177, 207, 209, 211, 220, 223
Eurasian 120
everyday life 101, 127, 181, 183, 208
evil eye 219, 220, 221, 222

families 3, 6, 31, 35, 36, 38, 74, 76, 77, 158, 175, 176, 199, 200, 201, 202, 207, 211, 213
family life 183
father 72, 76, 106, 114, 120, 159, 163, 190, 191, 192, 193, 196, 199, 206, 209, 210, 214, 216, 217, 223, 224, 225
fatherhood xiv, 4, 7, 192, 194, 198, 199, 201, 206, 207, 210, 217, 223, 224
fear(ed)(s) xiv, 13, 31, 33, 34, 35, 39, 42, 43, 51, 63, 64, 88, 90, 92, 94, 121, 163, 165, 170, 171, 172, 179, 180, 213
female allies 65
femininity xiii, 1, 36, 37, 48, 51, 54, 102, 125, 143, 156, 157, 158, 159, 160, 163, 164, 189, 232
feminism 4, 31, 34, 35, 36, 37, 42, 102, 155, 234
feminist anthropology 209
Filipino machismo 75, 77
Filipino masculinity 71, 72, 73

flying checkpoints 175, 178
friend(s) 24, 26, 35, 43, 47, 75, 76, 77, 78, 80, 81, 82, 87, 89, 90, 91, 94, 136, 140, 145, 159, 163, 164, 167, 183, 212, 213, 215, 217, 218, 222, 224, 225
frustration 22, 179, 181

gaijin smash 134, 136, 143, 148
gatarakwa 162, 164
gatekeeper(s)(ing) 13, 17, 61, 62, 63, 64, 65, 80, 91
gay(s) 14, 71, 73, 74, 76, 77, 78, 79, 81, 87, 138, 156, 159, 160, 165
gender identity 78, 114, 154, 156
gender performativity 78
gender socialisation 14, 17
gilo terminal 171, 182
Gusii 158, 159, 161, 162, 163, 164

hegemonic male xiv, 31, 32, 33, 34, 35, 37, 38, 39, 40, 41, 42, 44, 129, 156
hegemonic masculinity xiii, 41, 42, 72, 87, 93, 94, 95, 125, 126, 129, 160, 163, 189, 230, 231
hegemony 95, 126, 157, 159, 160
heritage 56, 106, 108, 114, 140, 155, 156, 210, 222
holism 90, 91, 95
homosexuality 51, 73, 78, 82, 83, 87, 160
hon'ne 136
humiliation 172, 175, 178, 182
husband(s) 20, 38, 72, 76, 92, 159, 163, 167, 183, 190, 192, 194, 195, 199, 200, 201, 206, 207, 214, 215, 216
husband-hood 198, 200, 201, 202

identity 31, 58, 64, 73, 76, 78–83, 86–89, 90, 93, 102, 106–107, 114, 118, 125–127, 130–131, 137, 155, 163–164, 206, 210–213, 216–217, 221, 223, 225
imagined communities 140
immobility 175
imposter syndrome 41, 42, 130
insider(s) 37, 51, 153, 164, 173
intifada 176, 178
invasions 176
Israel 172, 173, 176, 179, 180, 182
Israeli settlers 183
Italy 206, 207, 210, 212, 213, 214, 215, 216, 217, 218, 219, 220, 221, 224, 225

Japan 134, 135, 136, 137, 138, 139, 142, 144, 147, 148
Jeffrey's bay 61
Jericho 177, 178

Kikuyu 159, 162, 163, 164
kinship 74, 206, 207, 209, 210, 212, 213, 216, 217, 222, 223, 224
Kisii 154, 164, 165
Kyoto 135, 138, 144, 145

little wedges 198

M50 91
male fieldworker 87, 89, 93, 94, 95
male homosexuality 74
male privilege 14, 87, 92, 93, 95
maleness 2, 4, 50, 104, 111, 117, 118, 157, 189, 230
Mallapurāna 141
manhood 3, 15, 54, 58, 159
marriage 51, 78, 106, 145, 157, 158, 159, 211, 215, 216, 217
masculine holism 91, 95
masculine reflexivity 91
matrix of control 175
mentor(ing)(s) 14, 44, 117, 118
methods (research) 8, 18, 22, 25, 38, 85, 86, 92, 123 124, 138, 147, 207
Middle East 179, 180, 214
misogyny 13, 50, 72
mobility capital 23
moral anxiety 111
Mzungu 153, 155, 161, 162, 164, 167

narratives 18, 22, 27, 58, 64, 82, 146, 171, 175, 208
narrative turn 20
neoliberal labour market 140
neo-tribe(s) 52
New Zealand(ers) 26, 46, 47, 71, 78, 82, 108
non-western contexts 108, 172

objective researcher 87
obstacles 174, 179, 180, 182
occupation (Israeli) 22, 24, 25, 27, 62, 171, 172, 173, 175, 176, 178, 179, 181, 182, 183
ocean 52, 53

Orientalism 180
outsider 80, 110, 111, 114, 153

Padre Pio 206, 207, 210, 218, 221, 222
Palestinian society 182, 183
participant observation(s) 85, 207
patriarchal xiv, 7, 72, 74, 75, 81, 102, 104, 105, 106, 109, 111, 112, 113, 158
performance anxiety 82
persuasion 146
Pesaro 217, 218, 219, 222
Philippine society 72, 74
Pietrelcina 206, 210, 212, 213, 217, 218, 220, 221, 222
pilgrimage(s) 136, 206, 207, 208, 209, 210, 218, 222
politics 4, 58, 88, 94, 95, 109, 170, 171, 172, 180, 206, 216
positionality xiii, xiv, 23, 33, 37, 115, 117, 118, 119, 123, 124, 127, 129, 130, 134, 170, 173, 206, 207, 208, 211
prenatal classes 217
presences 189, 190, 199, 202, 203

queer anthropology 209, 223

race xiii, 14, 33, 87, 123, 130, 167, 170, 206, 214, 216, 217
racial politics 216
recreation(al) 6, 48, 49, 51, 56, 60
reflexivity xiv, 3, 4, 5, 7, 19, 83, 86, 88, 91, 92, 102, 103, 131, 172, 173, 190, 233
religion 26, 106, 134, 153, 165, 166, 167, 173, 210, 222
relocation(s) 7, 192, 194, 195, 201
research partner 117
restrictions 78, 170, 171, 172, 176, 177, 182
rights 35, 51, 143, 159, 160, 170, 214, 215, 216, 228
risk(ier)(s)(y) xiv, 5, 15, 48, 52, 64, 78, 94, 111, 120, 140, 146, 172
roles xiii, 3, 4, 14, 15, 48, 50, 51, 71, 72, 75, 76, 77, 82, 106, 110, 112, 114, 115, 116, 119, 120, 128, 157, 159, 190, 191, 195, 196, 197, 198, 200, 209, 232
rungu 166

safety xiv, 5, 76, 92, 165, 170, 175, 182, 217, 233
sailing 49, 52, 198

saints 220, 221, 222
Sanskrit 135, 136, 142, 146
scientific colonialism 95
security xiv, 24, 58, 94, 170, 171, 175, 176, 181, 182
Segregation Wall 175, 180, 183
self 7, 20, 21, 22, 42, 76, 78, 80, 82, 103, 104, 106, 107, 123, 129, 200, 223
sexism 47
sexual 32, 35, 39, 48, 54, 58, 64, 73, 77, 78, 115, 126, 157, 162, 164, 179, 209
sexual harassment 5, 148
sexuality 3, 6, 58, 72, 76, 77, 78, 82, 87, 89, 90, 95, 155, 172, 210
Shanghai 90
Singapore 87, 92, 116
social desirability bias 147
social facts 207
social networks 14, 209, 210, 213
social psychology 146
social media 41, 136, 139, 144, 145, 146, 147
scraped 147
sentiment analysis 147
Songzhuang 89
South Africa 57, 58, 60, 62
spacio-cidal 176
sponsor 116
sport(s) 33, 46, 48, 52, 53, 64, 215
status 3, 56, 58, 90, 95, 106, 114, 118, 119, 120, 140, 159, 163, 167, 196, 199, 209, 210, 211, 213, 214, 215, 217, 219, 223, 224

storytelling 20
student(s) 2, 33, 34, 37, 39, 41, 43, 57, 58, 89, 90, 114, 115, 116, 126, 135, 175, 177, 191, 214, 233
surf credentials 61, 62
surfing 49, 50, 52, 56, 60, 61, 62, 63, 64, 74, 75, 80, 81
surfing women 60

tatemae 136
teacher(s) 38, 47, 109, 116, 136, 138, 139, 140, 142, 143, 146, 162, 175, 198, 213
threat(s) 47, 64, 87, 143, 175, 179, 182
toughness 47, 48, 52
toxic masculinity xiv, 1, 3, 232, 233
traumatic 175
tribe(s) 2, 8, 140, 153
Triple Crisis 19, 157, 158, 161, 162, 164
trust 5, 40, 65, 80, 90, 95, 160, 161, 162, 163, 164, 208, 211, 219, 223, 224
typologising 18, 19

unequal power relation(s)(ship) 131, 179

well-being 3, 31, 40, 43, 170, 171, 175, 182, 217, 221, 223
wellness 134, 135, 136, 143, 146
West Bank 22, 23, 170, 173, 174, 175, 176, 177, 178, 179, 180, 182, 183
women-marriage 159

yoga industrial complex 136
Yogaland 134, 136, 140, 145, 146
yoga lifestyle(s) 141, 144